KU-566-769

Topics in Engineering
Volume 15
Edited by C.A. Brebbia and J.J. Connor

Solving Heat Radiation Problems Using the Boundary Element Method

Ryszard A. Bialecki

Computational Mechanics Publications
Southampton UK and Boston USA

Series Editors
C.A. Brebbia and J.J. Connor

Author
Ryszard A. Bialecki
Institute of Thermal Technology
Technical University of Silesia
44-101 Gliwice
Konarskiego 22
Poland

Published by

Computational Mechanics Publications
Ashurst Lodge, Ashurst, Southampton, SO4 2AA, UK
Tel: 44 (0)703 293223 Fax: 44 (0)703 292853
Email: CMI@uk.ac.rl.ib

For USA, Canada and Mexico

Computational Mechanics Inc
25 Bridge Street, Billerica, MA 01821, USA
Tel: 508 667 5841 Fax: 508 667 7582

British Library Cataloguing-in-Publication Data

A Catalogue record for this book is available
from the British Library

ISBN 1-85312-254-8 Computational Mechanics Publications, Southampton
ISBN 1-56252-178-0 Computational Mechanics Publications, Boston
ISSN 0952-5300 Series

Library of Congress Catalog Card Number 93-72438

Reprinted August 1995

Printed in Great Britain by Antony Rowe Ltd, Chippenham, Wiltshire

To the memory of my father

Contents

Preface

Scope and objectives

The book deals with a novel numerical method of treating heat radiation problems with special stress laid on coupling radiation with other modes of heat transfer. The material of the present work comprises heat radiation, heat conduction and the Boundary Element Method (BEM). The book is meant to be self contained. While making no pretense at completely covering these subjects the reader is supplied with a background sufficient to write a computer code capable of solving radiation problems of almost any degree of complexity.

The book is aimed at researchers and engineers working in the area of mathematical modelling of 'pure' heat transfer phenomena and related topics such as combustion, thermoelasticity, space engineering, etc. It should also be comprehensible to graduated students in physical engineering and applied mathematics having an interest, and some knowledge in numerical techniques and heat transfer.

The method of presentation is didactic, thus the more complex problems are approached by a stepwise addition of complicating factors. The author appeals to physical intuition rather than to the sound mathematical theory. Due to the interdisciplinary character of the material covered, the likelihood that the readership of this book will be unfamiliar, at least with some of the subjects dealt with, seems to be high. Therefore the material is written with the assumption that most of the readers will meet the subject for the first time. Hence, some introductory material concerning basic notions of heat radiation, heat conduction and the outline of the Weighted Residuals Method is included, both for the reader's convenience, and in order to introduce consistent notation.

As the majority of practical heat radiation problems encountered in engineering are three dimensional, the problems studied in this work are essentially in 3D. One dimensional problems, being nowadays only of historical importance, are not dealt with. Although the material on 2D problems spread throughout the work suffices to solve them, they are treated marginally.

A brief description of the book contents follows. No attempt is made to mention all topics covered, only the general thread of the development is indicated.

The initial part of Chapter 1 contains general characteristics of heat radiation problems and their inherent complexities. In the remaining portion of this chapter most frequently used discretization methods, i.e. Finite Difference Method (FDM),

Finite Element Method (FEM) and Boundary Element Method (BEM) are overviewed with stress put on their similarities and areas of efficient application.

The aim of Chapter 2 is to present a systematic derivation of the equations of radiative heat transfer. This is preceded by thorough definitions of all quantities needed to derive these equations. Two formulations of the integral equations of radiation are presented: the classic one, with a volume integral present, and a novel BEM formulation with the surface integral replacing the volume one.

The discretization process of BEM can be interpreted as a specific version of the Weighted Residuals Method (WRM). To follow the procedure of converting the original integral equations of heat radiation into sets of algebraic equations the knowledge of WRM is essential. The outline of this method is contained within Chapter 3. Simple numerical examples are solved in this chapter, in order to make the reader familiar with the practical aspects of WRM implementation.

Chapter 4 addresses the derivation of the integral equations of conduction. First, the governing differential equations with various types of associated boundary conditions are discussed. The original boundary value problem is transformed into an equivalent integral equation via WRM and integration by parts. As the stress in this book is laid on radiation, with the intention not to distract the attention of the reader by additional complexities, only steady state problems of linear and homogeneous material are considered. However, the allowance for nonlinear boundary conditions is included. This is done in order to take into account the coupling between heat conduction and radiation discussed in subsequent chapters.

The details of the BEM discretization technique applied to heat conduction are dealt with in Chapter 5. The procedure is described exhaustively, as the same algorithm is then used to handle the equation of radiation. Provided the material of this chapter is grasped, even a reader unfamiliar with BEM will readily understand the completely analogous discretization method employed in radiation problems. The chapter begins with basic notions of shape function approximation of both the unknown functions and the shape of the domain. Details of numerical integration over portions of the boundary (boundary elements) are dealt with in subsequent sections. The chapter also addresses some important aspects of the numerical implementation of BEM seldom described in the literature. This comprises specifically the question of adaptive integration. The experience gained by the present author shows that integration strongly influences both the accuracy of the results and the efficiency of computations, therefore one section of Chapter 5 has been devoted to this problem. The method the symmetry is dealt with in some BEM codes is also described. It is shown that the same technique can be employed to account for symmetry in heat radiation problems.

The question of discretization of the heat radiation equations in their BEM formulations is addressed in Chapter 6. First, the similarity of both the appearance and the order of kernel singularity of the integral equations of heat conduction and radiation is pointed out. Then the simple case of transparent medium filling a radiating enclosure is examined. More complex physical situations involving isothermal, nonisothermal, gray and nongray medium are discussed consecutively in the order of increasing complexity. The subsequent portion of the chapter is devoted to numerical evaluation of both regular and singular integrals over boundary elements. These integrals are necessary to compose the final form of the BEM matrices. The numerical examples included illustrate the flexibility of BEM and the relative ease with which various physical complexities can be taken into account. The closing part of the chapter contains discussion of the possible set of data needed to solve the radiation problem uniquely.

The book is organized in a fairly standard manner with the exception of Chapter 7 which is devoted to the overview of available techniques of solving heat radiation problems. Usually, this kind of material is discussed in the introductory part of the book, whereas in the present work it has been postponed until the novel BEM technique has been described in full detail. There are two reasons for such uncommon organization of the material. The first is that many excellent literature reviews on heat radiation are available in the literature [43, 73, 95], thus it is superfluous to repeat this work here. Instead, the material of Chapter 7 is meant to offer a deeper insight into the mutual relation of BEM to other techniques. This kind of analysis can be carried out in a reasonable way after the reader has gained sufficient knowledge about the BEM technique. The attention in the chapter is focused first on the zoning methods. It is shown that the classic view factor and Hottel's methods can be interpreted as a specific WRM solution of the standard integral equations of radiation and thus, they are closely related both to FEM and BEM. The difficulties arising when forming Hottel's zoning matrices are pointed out. The reason for these difficulties and an efficient way of how to circumvent them is described. A simple numerical example shows that BEM compares favourably with the zoning approach. Other groups of solution methods discussed in Chapter 7, and applicable to heat radiation problems, are the directional equation methods (flux, moments), Monte Carlo technique and some hybrid methods. Each group is analyzed in the context of its drawbacks, merits and recommended areas of applications and is compared with the BEM approach.

Chapter 8 examines the problem of coupling radiation with other heat transfer modes. The problems studied comprise coupling radiation with conduction in solid walls forming an enclosure filled by transparent or participating medium. The influence of heat convection is also examined briefly.

Finally, Chapter 9 contains some concluding remarks concerning the advantages and disadvantages of the BEM technique. The attention is focused on the comparison of BEM with the classic Hottel's method, being the most frequently used numerical technique of handling radiation problems. Both approaches are based on subdivision of the domain into zones (elements) and use numerical quadratures to compute the entries of the matrices. The advantages of BEM come primarily from the low cost of computing these entries. Hottel's method requires double integrals and volume integration, thus elements are computed by integration in four, five and six dimensions space. All entries of BEM matrices are computed by integration in two dimensions space, which considerably reduces the cost of discretization. As BEM is a novel method of attacking heat radiation problems, a fairly long list of open, or not satisfactorily clear problems, that can be topics of further research is included.

Part of the material included in this work contains a standard knowledge. This concerns the basic notion of heat conduction, radiation and the detailed description of BEM and its application to heat conduction. The remaining portion of the book arose from the material being a result of the (partially unpublished) research of the present author. As these topics are scattered throughout the book they will be highlighted here. Boundary only formulation of heat radiation equation, applying BEM to heat radiation in both participating and transparent media, applying BEM to coupled problems including radiation and conduction both in the presence and absence of participating media, as well as the interpretation of standard zoning approaches as a specific case of the Galerkin method are the most important examples of such topics. Usage of adaptive integration and treatment of symmetries in heat radiation are other novel aspects dealt with in this book.

Acknowledgments

It is a pleasure to record my indebtedness to Professor Jan Szargut of the Silesian Technical University of Gliwice who encouraged me to write this book, read the entire manuscript, and suggested several valuable changes to the text.

I benefited also from the helpful comments and corrections given by Professor Luiz Wrobel of the Wessex Institute of Technology who was kind enough to read the final version of the manuscript.

I wish to express my gratitude to my colleagues Dr. Andrzej J. Nowak, Prof. Kazimierz Kurpisz, Drs. Janusz Skorek and Adam Fic for many illuminating discussions and fruitful cooperation in the course of our association extending over many years at the Institute of Thermal Technology.

I feel indebted to Prof. Günther Kuhn of Erlangen University, Germany who provided me generously with the access to the computer facilities of Erlangen University and suggested some new topics of research. Dr. Rudolf Dallner also deserves my special thanks for his valuable hints concerning BEM computer implementation.

The author has received some support from the Polish National Science Foundation within the grant 3 1137 91 01. This help is gratefully acknowledged herewith.

Last but not least I wish to extend my sincere thanks to my wife Bożena and children Tomek and Zuzia who have suffered through long hours of neglect in the course of conducting the research underlying the book and later, when writing it.

Gliwice September, 1992

Ryszard A. Białecki

NOMENCLATURE

LATIN

a– absorption coefficient; $1/m$

$\mathbf{A}$– matrix in the discretized radiation equation

A_j– spatially dependent, unknown coefficients of the expression (7.20) approximating the intensity

b– radiosity defined as a sum of surface emission and energy reflected from an infinitesimal surface; $W \cdot m^{-2}$

$\mathbf{B}$– matrix in the discretized radiation equation

B_j– angular dependent, known coefficients of the expression (7.20) approximating the intensity

(c, d)– integration interval in integral equation (3.1)

C– known scalar in integral equation (3.1)

$\mathbf{C}$– matrix in the discretized radiation equation

C_1, C_2– constants in Planck's spectral energy distribution

C_3– constant in Wien's displacement law

d_l– length of ray within subsequent lth volume cell intersected by the ray

D– known scalar in integral equation (3.1)

D_ξ– maximum length of element in the direction of local coordinate ξ

$\mathbf{D}$– matrix in the discretized nonlinear conduction equation

e^r– radiative energy flux; $W \cdot m^{-2}$

e_b– blackbody emissive power; $W \cdot m^{-2}$

E– number of boundary elements

E^r– radiative energy; J

$f(x)$– solution of integral equation (3.1)

$\mathbf{f}$– vector accounting for internal heat generation and boundary conditions in the discretized equation of conduction

$g(y)$– known function in integral equation (3.1)

$\mathbf{g}$– known vector in WRM equations (3.7)

g_i– term due to internal heat generation in the discretized heat conduction equation

$\mathbf{G}$– matrix in the discretized equation of heat conduction

G_ξ– order of quadrature along local coordinate ξ

G_η– order of quadrature along local coordinate η

h– heat transfer coefficient; $W/(m^2 \cdot K)$

h_m– known functions defined by Eq. (3.4)
$\mathbf{H}$– matrix in the discretized heat conduction equation
i– irradiance defined as a sum of radiant energy incident on an infinitesimal surface; $W \cdot m^{-2}$
I– intensity of radiation; $W/(m^2 \cdot sr)$
I^e– number of nodal points assigned to boundary element ΔS_e
I_{rp}– number of volume cells intersected by a ray going from point $\mathbf{r}$ to $\mathbf{p}$
I_{ij}^{mn}– moment of radiation intensity
J_1, J_2– line integrals defined by Eqs. (6.31), (6.32)
J– number of terms in expression (7.20) approximating the intensity
k– heat conductivity; $W/(m \cdot K)$
$k(x, y)$– known kernel function of integral equation (3.1)
$\mathbf{K}$– matrix in the discretized BEM heat conduction equation
K, K_r, K_p, K_0– known kernel functions of integral equation of radiation
l_{x_i}, l_{x_j}– direction cosines with respect to coordinate x_i, x_j, respectively
L– number of volume cells in heat conducting medium
L_l– distance from the origin of the ray to the intersection with the subsequent lth volume cell passed by the ray
L_{min}– minimum distance between collocation point and boundary element
L_{rp}– line of sight
L_V– number of volume cells in participating medium
m_j– interpolating functions arising in the Dual Reciprocity Method
M– number of unknowns in WRM (3.2), (3.11)
n_{rx}, n_{ry}, n_{rz}– Cartesian coordinates of the unit normal at point $\mathbf{r}$
N– number of nodal points placed on the boundary
N_{rx}, N_{ry}, N_{rz}– Cartesian coordinates of the normal at point $\mathbf{r}$
$\mathbf{p}$– vector coordinates of observation (source) point
q– conductive heat flux normal to a surface; $W/(m^2)$
$\mathbf{q}$– vector of nodal conductive heat fluxes
q^r– radiative heat flux defined as a flux of radiant energy gained by an infinitesimal surface; $W/(m^2)$
$\mathbf{q}^r$– vector of nodal radiative heat fluxes
q_x– x component of the vector of conductive heat flux
q_V– heat source defined as an amount of energy generated within an elemental volume per unit time; $W/(m^3)$
$\mathbf{q}_V$– vector of nodal heat sources caused by phenomena other than radiation
q_V^r– radiative heat source defined as an amount of radiant energy gained by an elemental volume per unit time; $W/(m^3)$
$\mathbf{q}_V^r$– vector of nodal radiative heat sources

$r(y)$– residual of integral equation (3.1) defined by Eq. (3.5)

r– vector coordinates of current (integration) point

R– vector being a difference between the current point **r** and observation point **p**; $\mathbf{R} = \mathbf{r} - \mathbf{p}$

R', R''– distance between the point of ray origin **r** and current points $\mathbf{r}'$, $\mathbf{r}''$ placed on the line of sight, respectively

R_x, R_y, R_z– coordinates of vector **R** in Cartesian coordinates system

s– order of integrand singularity

S– surface area; m^2

t– time; s

t– unknowns of the discretized heat conduction BEM equations- temperatures or heat fluxes at nodes located on the boundary

T– temperature; K

$\hat{T}_j$– functions associated by Eq. (4.26) with interpolating functions m_j arising in the Dual Reciprocity Method

T– vector of nodal temperatures

U– number of spectral intervals

v_i– weighting factor in the weighted sum of gray gases model

V– volume

$w_j(x)$– weighting function in the WRM technique

w_n– weight of Gaussian quadrature

W– known matrix in WRM equation (3.7)

GREEK

α– absorptivity

$\boldsymbol{\alpha}$– vector of unknowns in the WRM equation (3.7)

α_m– degrees of freedom sought for in WRM

β– shadow zone function

$\delta(x)$– Dirac's distribution

δ_{ij}– Kronecker's symbol

ΔS– boundary element

ϵ– emissivity

ε– integration accuracy

η– local coordinate within boundary element

θ– azimuth angle

Θ– 3D shape function approximating the variation of a function

κ– optical thickness

λ– wavelength; m

ξ– local coordinate within boundary element

σ– Stefan–Boltzmann constant

$\tau(\mathbf{r},\mathbf{p})$– transmissivity

ϕ– angle between a given direction and a normal to the surface

Φ– shape function approximating the variation of a function

Ψ– shape function approximating the geometry

ρ– reflexivity

Ω– solid angle; *sr*

INDICES

subscripts

b– blackbody

e– boundary element

p– observation (source) point

r– integration (current) point

$\perp$– normal to the surface

$0-\lambda$– in spectral range $0-\lambda$

$\lambda_1-\lambda_2$– in spectral range $\lambda_1-\lambda_2$

ϕ– directional quantity, ϕ is an angle measured with respect to the surface normal

λ– spectral

superscripts

E– emitted radiation

e– boundary element

f– fluid exchanging heat by convection with a solid surface

G– result of Gaussian pre-elimination

i– incoming radiation

L– nonradiating surface-linear boundary conditions

m– participating medium

o– outgoing radiation (sum of emitted and reflected energy)

r– radiative energy

R– radiating surface

u– spectral interval

SPECIAL

$\mathcal{M}$– symmetry transformation

∇– gradient differential operator

∇^2– Laplace differential operator

$|\mathbf{r}-\mathbf{p}|$– distance between points **r** and **p**

$\frac{\partial}{\partial n}$– differentiation along the outward normal

$\int_{L_{rp}}(\,)\,dL_{rp}$– integration along a line of sight linking points **r** and **p**

Chapter 1

Introduction

1.1 Characteristic features of heat radiation problems

All bodies at temperatures above absolute zero emit energy in a form of electromagnetic waves. A portion of this energy flux when impinging other bodies is absorbed. As a result, net energy flow occurs from a body of higher temperature to a body having lower temperature. This mode of energy transfer is termed *heat radiation.* The wavelength range encompassed by thermal radiation is approximately 0.1 to 100 μm . Heat radiation is, as each wave propagation phenomenon, of dual nature. It possesses the continuity properties of electromagnetic waves and the corpuscular properties characteristic of photons.

Radiation plays a dominant role in energy transfer at elevated temperatures and in the presence of rarefied gases. The amount of heat transported by radiation in industrial furnaces and combustion chambers typically reaches 90%. Heat exchange in space and solar heating devices is 100% due to heat radiation. Even at relatively low temperatures characteristic of central heating systems, nearly half of the heat is transferred by radiation. Thus, radiative heat analysis constitutes the crucial portion of the calculation of temperature fields in various branches of science and technology.

Owing to the progress in computer technology, mathematical modelling has become a cheap and reliable tool of engineering design. Almost all phenomena that modellers deal with are governed by differential equations. There are many well established numerical techniques for solving differential equations of mathematical models.

Radiation is one of few phenomena governed by an integral equation. This feature is a source of both conceptual and computational difficulties to most engineers whose mathematical backgrounds are based on differential equations. Additional complexities inherent in heat radiation computations result from the severe nonlinearity and very complex characteristics of the material properties appearing in the radiation transport equations. The differences between radiation and other heat transfer modes will now be discussed briefly.

First the physical consequences arising from the nature of the integral equations will be pointed out. Consider a point lying on a boundary of an enclosure formed by solid walls. The temperature field within the solid walls is obtained upon solving a differential equation. The conductive heat flux is then obtained by differentiation of the temperature field at that very point. Thus, the conductive heat flux depends solely on temperatures lying in the close vicinity of the point under consideration. Radiative heat flux gained by a point lying on the concave surface of the solid is obtained upon solving an integral equation. This means that the radiative flux depends on all temperatures of this surface. Contrary to the case of heat conduction, temperatures at points laying far from the considered point can significantly influence the heat flux at that point. Moreover, temperatures in the nearest vicinity of the point under consideration often do not exert any influence on the radiative flux at this point.

Because radiation is transported via electromagnetic waves, it can be transferred even in a vacuum. Other heat transfer modes, i.e. conduction and convection, require a physical medium for heat interchange to occur.

The simplest possible case of radiative heat exchange is two parallel isothermal black surfaces separated by a transparent medium. The *Stefan-Boltzmann law* states that in this case the heat flux is proportional to the difference of the fourth powers of the surface temperatures. In some practical situations the order of the nonlinearity can be even higher. This is in contrast with conductive heat transfer where the heat flux is proportional to the temperature gradient. Similarly, convective heat flux is approximately proportional to the difference of the surface and bulk fluid temperature.

Typically, material properties entering the equations of radiative transfer depend strongly on the length of the electromagnetic wave. In the case of gases this dependence assumes a very complex form, arising from quantum mechanics. Gases can be transparent to radiation in certain intervals of the spectrum while participating in radiative heat transfer in other intervals. It is known that carbon dioxide present in the atmosphere is practically transparent to sun radiation while it absorbs the radiation emitted by the earth. The *greenhouse effect* causing dangerous changes in the global climate is attributed to the increase of the concentration of CO_2 in the atmosphere.

A characteristic feature of radiation is that it can be transferred directly from one location to another only when one point can be 'seen' when looking from another, i.e. it does not lay in a *shadow zone*. The presence of shadow zones should be taken into account in heat radiation calculations. This leads to complex algorithms and long computing times.

1.2 Discretization methods

Typical mathematical models are formulated in terms of differential equations. The cases when these equations can be solved using analytical techniques are limited to very simple situations. The bulk of practical problems are solved using numerical methods. The common feature of numerical techniques is the discretization of the problem, i.e. transforming the governing equations to a system of algebraic equations. Three discretization methods among many other available techniques have passed into common usage: *Finite Difference Method*, *Finite Element Method* and *Boundary Element Method*. The popularity of these methods is due to the ease of building, based on these techniques, a general purpose software capable of handling problems in bodies of arbitrary geometry and with any combination of boundary and initial conditions. Options enabling one to deal with practically any kind of nonlinearity can be readily included in these codes.

In the Finite Difference Method (FDM) the derivatives are replaced by finite differences of unknown function values at selected nodal points. Alternatively, the equations of FDM can be derived using conservation laws. The set of algebraic equations resulting from the discretization has a characteristic band structure, i.e. the nonzero entries of the matrix are gathered in the vicinity of the diagonal of the matrix. The set is solved for unknown nodal function values.

Equations of the Finite Element Method (FEM) [97] are obtained upon employing variational principles, e.g. minimum energy, maximum entropy production. An alternative, somewhat more general approach of discretizing the governing equations, is to apply the weighted residuals technique (see Chapter 3). A characteristic feature of the FEM is the usage of locally based interpolation functions termed *shape functions* to approximate the shape of the body and the distribution of the unknown function. The resulting equations set has a similar band structure as in the case of FDM. Attempts were reported in the literature of using the power and versatility of FEM to solve radiation problems [69, 70]. Contrary to other fields of application, employing FEM in heat radiation analysis has not become popular. The reason for this is that FEM has been developed to handle differential equations. Using FEM to solve integral equations is cumbersome.

The Boundary Element Method (BEM) [4, 22, 23] is the youngest general purpose numerical method. Details of BEM applied to heat conduction will be discussed in Chapter 4. Here, only the characteristic features of this technique distinguishing it from FDM and FEM will be mentioned. BEM relies on transforming the original differential equation into an equivalent integral one. This is accomplished using either Green's identity or the reciprocity theorem.

The integral equation of BEM is formulated usually only on the boundary of the domain, hence only the boundary needs to be discretized. It is equivalent to a reduction of the dimensionality of the problem by one. This reduction leads to substantial time economy in both data preparation and computing, and thus it is considered the main advantage of BEM. The typical discretization method used in BEM is the nodal collocation. Similarly as in FEM, shape function interpolating both the boundary and the unknown function are employed. The matrices arising in BEM are fully populated. The general scheme of BEM is shown in Fig. 1.1.

As the governing equation of radiation is an integral one, the idea of using BEM for the solution of radiation problems naturally arises. Employing BEM for the solution of heat radiation problems has been addressed in the literature [6, 7, 8, 9, 10, 11, 12, 20]. The technique proved to be efficient and easy to implement. The present book is a generalization and continuation of the aforementioned works. Reference [62] contains a review of the works published before 1989.

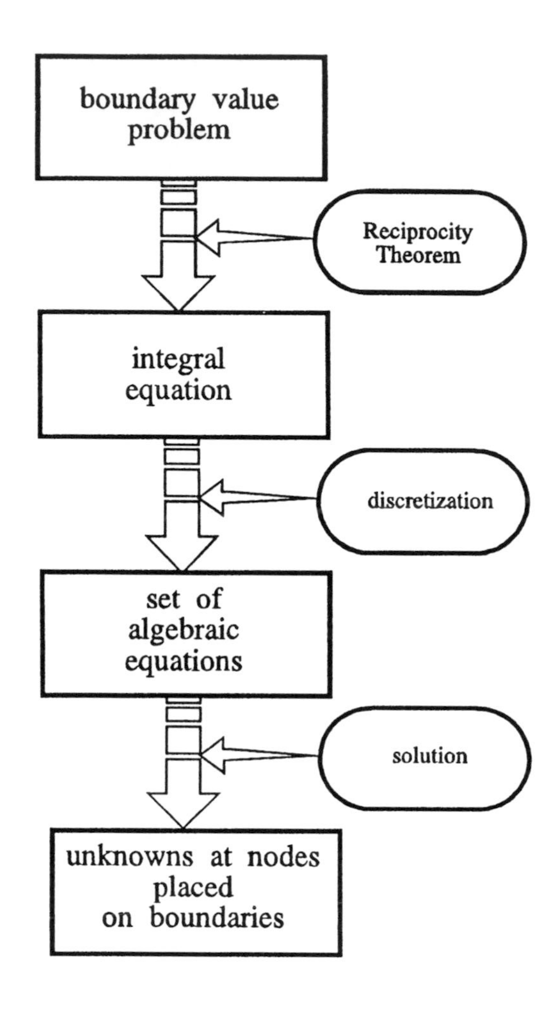

Figure 1.1: General scheme of the Boundary Element Method.

Chapter 2

Governing equations of heat radiation

Within this chapter only basic notions used in the subsequent sections dealing with numerical treatment of radiation heat transfer problems will be introduced. Extensive discussion of the physics of heat radiation can be found in monographs devoted to radiative analysis [64, 77, 82].

2.1 Basic notions

2.1.1 Blackbody radiation

Each body emits radiative energy. *Radiative energy flux* is defined as the amount of radiative energy passing a unit surface in a unit time:

$$e^r = \frac{d^2 E^r}{dS\, dt} \tag{2.1}$$

where:

e^r– radiative energy flux; $W \cdot m^{-2}$,

E^r– radiative energy; J,

dS– infinitesimal surface; m^2,

dt– infinitesimal time increment; s.

Radiation is transferred via electromagnetic waves whose lengths (spectrum) range from 0 to ∞. The amount of energy transported at a given wavelength is a function of this length. Functions associated with transfer at a certain wavelength are referred to as *spectral* or *monochromatic quantities*. Hereafter, these quantities will be denoted by appending a subscript λ to the appropriate symbol. Functions associated with the entire spectrum are referred to as *total* or *panchromatic quantities*. Let A_λ denote arbitrary spectral quantity and A stand for its panchromatic (total) counterpart. Due to the aforementioned notation convention these two functions are interrelated by the relationships

$$A_\lambda = \frac{dA}{d\lambda} \tag{2.2}$$

$$A = \int_0^\infty A_\lambda \, d\lambda \tag{2.3}$$

To simplify the notation index λ will be skipped in all cases where no confusion can arise. Equations where the index λ is explicitly written are to be interpreted as valid only for spectral quantities. Other equations can be interpreted as written in terms of either total or spectral quantities. For such equations their spectral counterpart can be obtained from the total one upon appending the index λ to appropriate symbols.

Transport of radiation takes place along straight lines referred to as *lines of sight.* To describe the transfer of heat radiation along a line of sight a notion of *intensity of radiation* is introduced. Intensity of radiation is defined as a radiative flux passing through an infinitesimal surface orthogonal to the line of sight and subtended within an infinitesimal solid angle centered around the line of sight. This definition can be expressed as (see Fig.2.1)

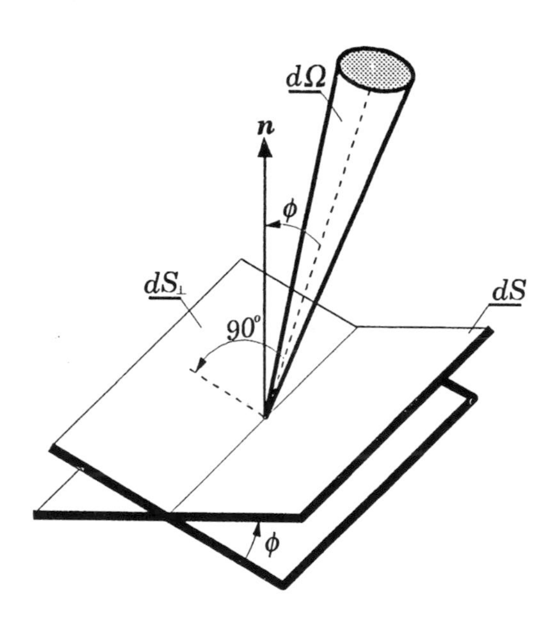

Figure 2.1: Notation pertinent to the definition of intensity of radiation.

$$I = \frac{d^3 E^r}{dS_\perp \, dt \, d\Omega} \tag{2.4}$$

where:

I– intensity of radiation; $W/m^2 \cdot sr$),

$d\Omega$– differential solid angle; sr,

$dS_\perp$– differential surface normal to the line of sight; $dS_\perp = dS \cos\phi$,

ϕ– angle with which the line of sight is inclined with respect to the surface normal.

The intensity is connected with the energy flux and hence, it can be interpreted as a vector having a direction of the line of sight. As an infinite number of lines of sight cross at a given point, an infinite number of intensities can be assigned to a chosen point.

A perfect emitter of radiant energy is called a *blackbody*. From all bodies of the same temperature a blackbody emits maximum energy. *Kirchhoff's law* states that a blackbody is also a perfect absorber of radiant energy, thus it absorbs the entire incident radiative energy. Quantities associated with this reference body will be marked hereafter by appending a subscript b to appropriate symbols.

A blackbody, being by definition an ideal emitter of radiation, emits energy uniformly in all directions. Hence, the *intensity of blackbody radiation* I_b is independent of direction. This property of a blackbody is referred to as the *isotropy of blackbody emission.*

The flux of radiative energy emitted by an elemental blackbody surface is termed *blackbody emissive power*. This flux can be computed upon integrating the normal component of the intensity vectors over the entire hemisphere centered at that surface

$$e_b = \int_{2\pi} I_b \cos\phi \, d\Omega \tag{2.5}$$

where:

I_b– blackbody intensity of radiation,

e_b– blackbody emissive power.

Referring to Fig. 2.2 a differential solid angle can be related to the polar angle ϕ and the azimuth angle θ by

$$d\Omega = \sin\phi \, d\phi \, d\theta \tag{2.6}$$

Taking into account Eq. (2.6) and performing appropriate integration Eq. (2.5) yields a relationship linking the intensity and the emissive power of the blackbody

$$e_b = I_b \int_{\Theta=0}^{2\pi} \int_{\phi=0}^{\pi/2} \cos\phi \, \sin\phi \, d\phi \, d\theta = \pi I_b \tag{2.7}$$

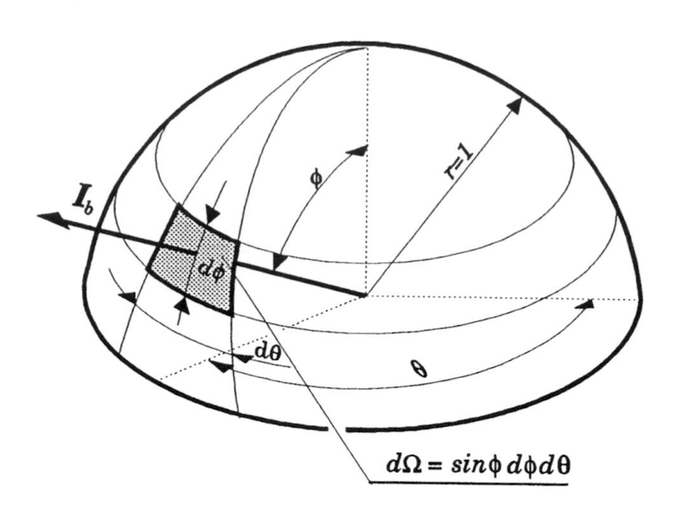

Figure 2.2: Unit hemisphere used to obtain the relation between blackbody intensity and emissive power.

Quantum mechanics yields an equation expressing the blackbody spectral emissive power as a function of temperature and wavelength. This relationship is known as the *Planck's function* and has the form

$$e_{b\lambda} = \frac{2\pi\, C_1}{\lambda^5 \left[\exp\left(\dfrac{C_2}{\lambda T}\right) - 1\right]} \tag{2.8}$$

where:

λ– wavelength; m,

T– temperature; K,

C_1– constant; $C_1 = 0.59544 \cdot 10^{-16}$; $W \cdot m^2$,

C_2– constant; $C_2 = 1.4388 \cdot 10^{-2}$; $m \cdot K$.

Strictly speaking Planck's function also depends on the *refraction index* of the medium. This index is defined as the ratio of the speed of light in a vacuum and in the medium. The refraction index of gases is very close to 1. For the sake of simplicity, radiation transfer within media having a refraction index equal to 1 will be considered throughout this book.

The plot of the spectral emissive power is shown in Fig. 2.3. It can be seen that at a certain wavelength the function assumes maximum. Vanishing of the derivative of the Planck's function (2.8) with respect to wavelength yields the location of this

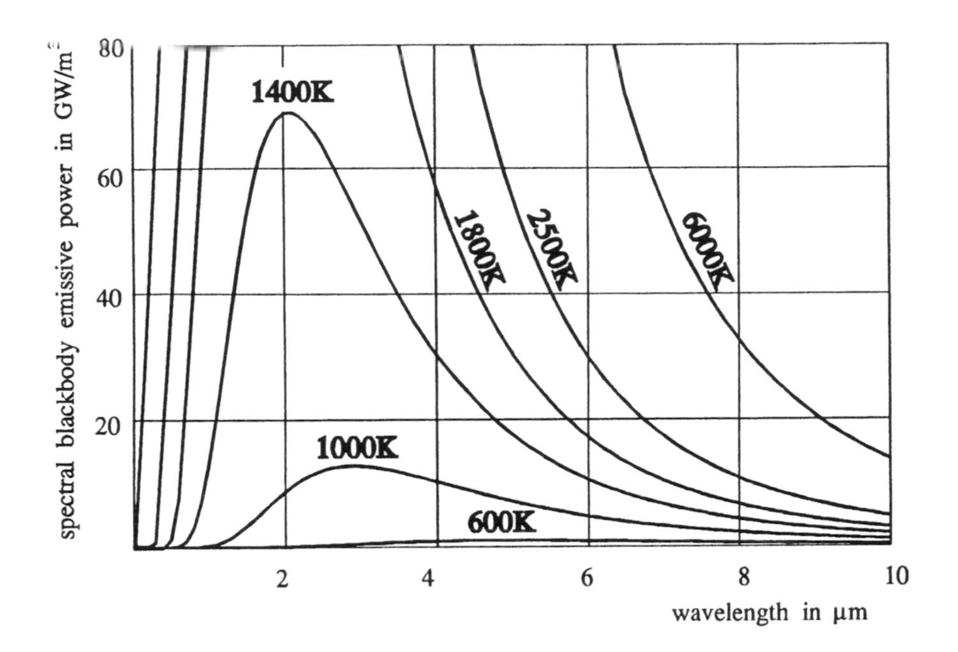

Figure 2.3: Blackbody spectral emissive power as a function of wavelengths for different temperatures.

maximum. Appropriate formula is known in the literature as the *Wien's displacement law*

$$\lambda_{max} T = C_3 \tag{2.9}$$

where λ_{max} is the wavelength corresponding to a maximum emission of radiation and $C_3 = 2.8978 \cdot 10^{-3}\ m \cdot K$.

Inspection of $e_{b\lambda}$ curves in Fig 2.3 shows that at elevated temperatures they decay rapidly to zero in the vicinity of their maxima. Moreover, the higher the temperature of the blackbody, the steeper is the curve. The dominant fraction of the radiation is hence transmitted in the small wavelength interval containing the maximum of spectral emissive power and the width of this interval decreases with the increase of temperature.

The fraction of blackbody emission contained within an arbitrary wavelength extending from 0 to an arbitrary wavelength λ can be calculated from Wiebelt's approximations [77]

$$e_{b\,0-\lambda} = \frac{15}{\pi^4} \sum_{m=1,2,\ldots} \frac{e^{-mv}}{m^4} \left\{ \left[(mv+3)mv+6 \right] mv + 6 \right\} \ ; \ v \geq 2 \tag{2.10}$$

$$e_{b\,0-\lambda} = 1 - \frac{15}{\pi^4} v^3 \left(\frac{1}{3} - \frac{v}{8} + \frac{v^2}{60} - \frac{v^4}{5040} + \frac{v^6}{272\,160} - \frac{v^8}{13\,305\,600} \right) \ ; \ v < 2 \tag{2.11}$$

where $v = C_2/(\lambda T)$ and

$$e_{b0-\lambda} = \int_0^{\lambda} e_{b\lambda}\, d\lambda \tag{2.12}$$

Energy emitted by a blackbody within a wavelength range $[\lambda_1 - \lambda_2]$ can be computed as

$$e_{b\lambda_1-\lambda_2} = e_{b0-\lambda_2} - e_{b0-\lambda_1} \tag{2.13}$$

The energy emitted by a unit blackbody surface in a unit time within the entire spectrum can be computed from a relationship known as the *Stefan-Boltzmann law*

$$e_b = \int_0^{\infty} e_{b\lambda}\, d\lambda = \sigma T^4 \tag{2.14}$$

where σ denotes the Stefan-Boltzmann constant; $\sigma = 5.669 \cdot 10^{-8}\ W/(m^2 \cdot K^4)$

2.1.2 Radiation of real surfaces

Radiation can be absorbed by a medium it traverses. Certain materials (and a vacuum) do not interact with the radiation. Such media are termed *transparent*. Remaining media are said to be *participating*. In this case radiation is absorbed in a very thin layer in the vicinity of the surface of the medium (typically $1\mu m$ to $1\ mm$) and the material is termed an *opaque medium*. For such media radiation can be regarded as taking place solely on their surface. Most solids can be treated as opaque substances.

Real substances differ from the ideal blackbody model because at the same temperature they emit less radiative energy. The fraction of blackbody emission radiated by a real surface of an opaque body is termed *emissivity* and is denoted as ϵ. Similarly, a real surface absorbs only a fraction of the incident radiant energy. This fraction is called *absorptivity* and is denoted as α. From *Kirchhoff's law* it follows that

$$\epsilon = \alpha \tag{2.15}$$

Obviously, incident radiation not absorbed on an opaque surface must be reflected therefrom. The fraction of incident energy reflected from a surface is termed *reflexivity* and is denoted as ρ. Energy balance yields a simple relationship valid for total and spectral quantities

$$\rho = 1 - \alpha = 1 - \epsilon \tag{2.16}$$

Emissivity, absorptivity and reflexivity are material properties characteristic of a surface. Typically, these properties vary within the spectrum. Surfaces whose radiative properties are independent of the wavelength are termed *gray*. Equations (2.15) and (2.16) concern such surfaces. Their spectral counterpart can be readily obtained upon appending a subscript λ to all symbols.

All mentioned material properties of the surface depend mainly on its roughness. Impurities, temperature and thin layer coating can also influence emissivity and reflexivity. For certain cases these properties can be calculated from electromagnetic theory. The final relationships link radiative properties with electrical and optical ones [77]. However, the predictions of properties obtained theoretically are inaccurate. Thus, radiative properties of real surfaces can be found only by carrying out appropriate measurements. Extensive tabulation of the measured data can be found in the literature [72].

Material properties of real surfaces can vary with the direction of the ray incident upon and leaving a surface. Angular dependence of emissivity and absorptivity has been observed, but for most practical cases this dependence can be neglected. Reflection from real surfaces can strongly depend on the direction of the incoming ray. There are two limiting models for the angular dependence of the reflexivity. The first one is called the *specular* or *mirrorlike reflection* . For surfaces characterized with this type of reflection, the normal to the surface, as well as the outgoing and incoming rays, are coplanar. The outgoing and incoming rays are placed on opposite sides of the normal and are inclined at the same angle to it. The second limiting case is a *diffuse reflection* where the direction of the outgoing ray is independent from the direction of the incoming one. The reflection from real surfaces can be regarded as a weighted sum of these two models. It should be stressed that there is not enough reliable data concerning the angular dependence of the radiative properties available in the literature. Therefore, unless explicitly otherwise stated, only properties independent on the direction, i.e. *diffuse emission, absorption*, and *reflection* , will be considered hereafter.

Let I_ϕ^E stand for the intensity of radiation emitted in a direction inclined by an angle ϕ to the surface normal. From the assumption of diffuse emission (ϵ does not depend on the direction) and isotropy of blackbody radiation it follows that

$$I_\phi^E = \epsilon \, I_b \tag{2.17}$$

Radiation outgoing from a surface is a sum of energy emitted by this surface and reflected therefrom. The integral of this sum over all admissible directions, is referred to as the *radiosity* and is denoted by b. Hence, radiosity can be defined as the radiant energy flux streaming away from an infinitesimal surface. The radiant energy flux incident upon an infinitesimal surface from all admissible directions is termed *irradiance* and is denoted by i. For the sake of simplicity it will be assumed hereafter that surface at the observation point is smooth. Hence, the admisible directions extend over the entire hemisphere, i.e. solid angle 2π). Observation points at corner points require special treatment [36].

Consider an elemental surface impinged by rays incoming from all directions. It is obvious that only normal components of the incoming intensity vectors contribute to the energy absorbed by the surface. Thus, with ϕ standing for the angle made by the incoming ray and the normal to the surface, the irradiatiation can be written as

$$i = \int_{2\pi} I_\phi^i \cos\phi \, d\Omega \tag{2.18}$$

where I_ϕ^i stands for the radiation intensity incident on the infinitesimal surface from the direction inclined by ϕ to the normal.

For diffusively reflecting and emitting surfaces, radiosity and irradiance are connected by a simple heat balance over the infinitesimal surface (see Fig. 2.4)

$$b = \epsilon e_b + \rho i \tag{2.19}$$

The amount of radiant energy reaching a surface and leaving it are not equal. The net radiant energy flux gained by an elemental surface is termed the *radiative heat flux* and is denoted by q^r. To maintain thermal equilibrium this amount of energy should be carried away by other modes of heat transfer, i.e. by conduction and/or convection. Thus, radiative heat flux enters the boundary conditions associated with the differential equations of conduction and convection. From the definition of the radiative heat flux it follows that it is calculable as a difference between the absorbed, incoming radiation and the emitted radiation. Consider an arbitrary direction inclined by an angle ϕ to the surface normal (see Fig. 2.5).

The portion of energy incoming within an elemental solid angle centered about this direction and absorbed on the surface is $\epsilon\, I^i_\phi \cos\phi$. The surface emits within the same solid angle an amount of energy equal to $\epsilon\, I_b \cos\phi$. The difference of these two energy fluxes integrated over the entire hemisphere yields the radiative heat flux

$$q^r = \epsilon \int_{2\pi} \left(I^i_\phi - I_b\right) \cos\phi \, d\Omega \tag{2.20}$$

Taking into account the definition of the irradiance [Eq. (2.18)] and that of black-body emissive power [Eq. (2.5)] Eq. (2.20) can be rewritten in the form (see Fig. 2.6)

$$q^r = \epsilon\, i - \epsilon\, e_b \tag{2.21}$$

Equation (2.21) can be used to eliminate the irradiance from the definition of radiosity (2.19). With Eq. (2.16) used to eliminate the reflexivity one arrives at

$$b = e_b + \frac{1-\epsilon}{\epsilon} q^r \tag{2.22}$$

Let I^o_ϕ denote the directional outgoing intensity, i.e. radiation intensity leaving the surface in a direction inclined by an angle ϕ to the normal. The superscript o denotes here the outgoing intensity, as opposed to the incoming denoted by the superscript i and emitted one marked by the superscript E. Noting that the radiosity is the entire radiative energy flux leaving the surface, the outgoing intensity associated with an elemental solid angle can be expressed as

$$db = I^o_\phi \cos\phi \, d\Omega \tag{2.23}$$

In accordance with the assumed diffuse distribution of both emitted and reflected radiation, the outgoing intensity is direction independent. Integration of Eq. (2.23) over the solid angle 2π compare with Eqs. (2.6), (2.7)] produces

$$b = \pi I^o_\phi \tag{2.24}$$

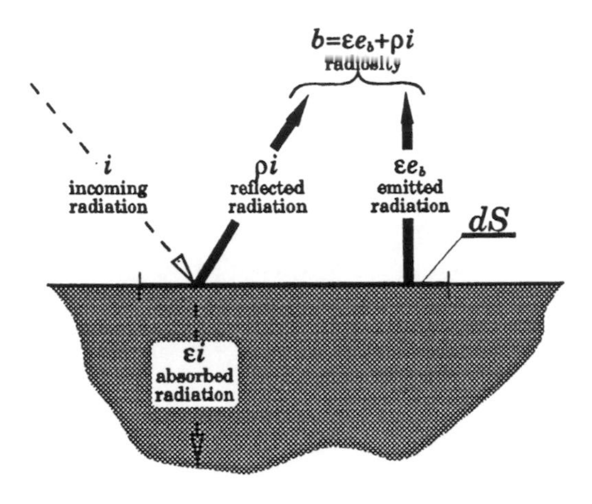

Figure 2.4: Definition of radiosity.

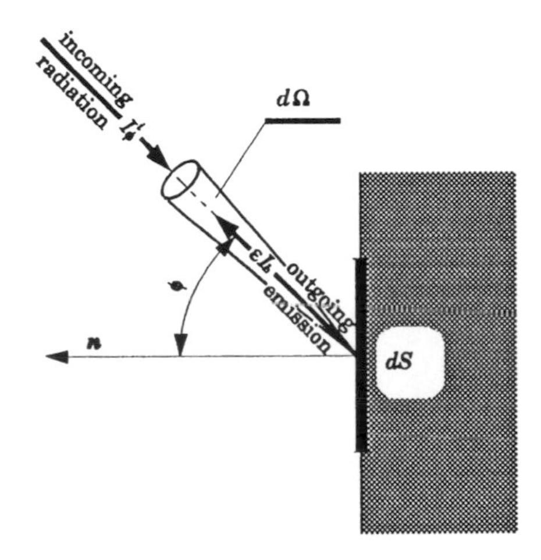

Figure 2.5: Radiative energy balance on an infinitesimal surface.

Finally, the intensity of radiation leaving the surface can be written in terms of emissive power and radiative heat flux. Combination of Eqs. (2.22) and (2.24) yields

$$I_\phi^o = \frac{1}{\pi}\left(e_b + \frac{1-\epsilon}{\epsilon}q^r\right) \tag{2.25}$$

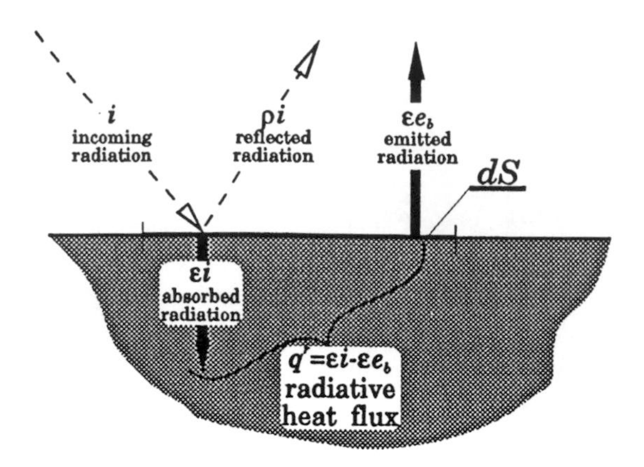

Figure 2.6: Definition of radiative heat flux.

2.1.3 Radiant energy transfer through participating media

As already mentioned for most solids, radiation can be considered as a surface phenomenon. The depth of radiation penetration of gases and liquids and some solids, e.g. glass, silicon can be large. Such substances are termed *semitransparent.* Radiation within semitransparent media can be viewed as a volumetric phenomenon.

When travelling through a semitransparent material radiation can be *attenuated* and *scattered.* Attenuation causes the intensity of radiation to decrease. The direction of the radiation however, remains unaltered. Scattering takes place when the ray is redirected. This can occur when a ray impinges a small particle, e.g. dust or fog droplet.

Attenuation and scattering diminish the intensity of radiation. An increase of intensity is caused both by the emission of the medium traversed by the ray and the inscattered rays. Including scattering in the model of heat radiation transfer leads to immense mathematical and numerical difficulties. In most engineering applications the influence of the scattering mechanism onto the overall radiative transfer can be neglected. It will be assumed throughout this book, that scattering is negligible. Such a model of radiation transfer in participating media is referred to in the literature as the *emitting-absorbing medium* approximation.

Participating media differ in their capability to emit and absorb radiation. From *Kirchhoff's law* it follows that at a given point within the medium the capability of absorbing and emitting radiation must be equal. The material property describing this capability is termed the *absorption coefficient* and will be denoted by a while its spectral analog is given by a_λ.

Absorption of radiation within a participating medium causes a decrease of the intensity of the ray travelling through the medium. On the other hand emission of radiant energy by the molecules of the medium increases the intensity of radiation.

Consider a pencil of rays travelling from point **r** to **p** along a line of sight. The increase of radiation intensity taking place as the radiation passes an infinitesimal path along the line of sight can be described by a differential equation

$$\frac{dI}{dL_{rp}} = -a\left[I - I_b\left(T^m\right)\right] \tag{2.26}$$

where dL_{rp} is the infinitesimal path along the line of sight and T^m denotes the temperature of the medium.

Equation (2.26) is termed the *differential form of the directional equation of radiative transfer.*

The radiative energy gained by an infinitesimal volume element is referred to as the *radiative heat source* and is denoted by q_V^r. It can be computed upon summing up the decrease of intensity defined by Eq. (2.26) associated with all lines of sight passing the element. This is accomplished by integrating Eq. (2.26) over all admissive directions, i.e. over the solid angle 4π (see Fig. 2.7).

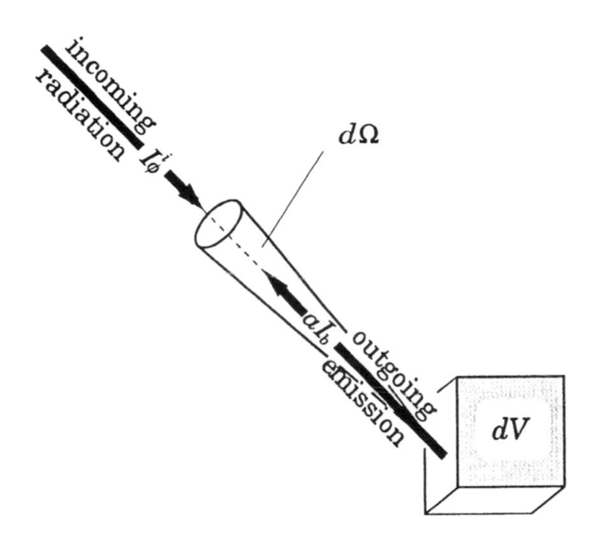

Figure 2.7: Radiative energy balance on an infinitesimal volume.

The result reads

$$q_V^r = -\int_{4\pi} \frac{dI}{dL_{rp}}\, d\Omega = a \int_{4\pi} \left[I_\phi^i - I_b(T^m)\right]\, d\Omega \tag{2.27}$$

To maintain thermal equilibrium the gain of radiative energy of an elemental volume must be carried away by other modes of heat transfer. Thus the function

q_V^r contributes to the source term of the conductive or convective energy transfer equations.

Contrary to radiative surface properties, i.e. absorptivity, emissivity and reflexivity, the absorption coefficient depends strongly on the wavelengths of the radiation. Few existing media have an absorption coefficient that can be regarded as constant within the entire spectrum. Such materials are termed *gray media.* Most semitransparent media belong to the group of *nongray media.* The absorption coefficient of such media varies substantially with the wavelength of radiation. Typical examples of such media are gases. Within certain spectral intervals called *bands* gases interact with the radiation. The remaining portion of the spectrum is covered by *windows.* Within windows gases are transparent to radiation, i.e. they do not interact with the radiation travelling through them.

In the process of absorption and emission of radiation quantum effects are of importance. Absorption of a quantum of energy by a gas molecule is associated with a peak of the absorption coefficient curve. Such a peak is termed a *spectral line.* Each spectral band contains many closely spaced spectral lines. The spectral lines can overlap, some effects, (e.g. pressure) influence the shape of a single line. All of these phenomena cause the dependence of the absorption coefficient on wavelength to be described by a very irregular curve (see Fig. 2.8). For the sake of simplicity the absorption coefficients are usually approximated by simple functions within spectral bands. There is a vast amount of literature on *equivalent band absorption coefficients* and the *bandwidth,* defined as the spectral interval containing the major absorption capability of a given band [25, 29, 30].

2.2 Directional equation of radiative transfer

Let $I(\mathbf{r})$ denote the intensity at the point where the ray originates. This quantity can be interpreted as an initial condition associated with the differential equation (2.26). Let $\mathbf{r}'$ stand for a current point lying on the line of sight and $I_b[T^m(\mathbf{r}')]$ denote the intensity of radiation of a blackbody having a temperature of the medium at point $\mathbf{r}'$ (see Fig. 2.9).

Incoming intensity at observation point $\mathbf{p}$ associated with the direction of the line of sight can be obtained upon integrating Eq. (2.26) formally along that line. Let the point $\mathbf{r}$ where the ray originates be placed on a solid surface bounding the participating medium. It will be assumed that the surface behaves like a diffuse emitter and reflector of radiant energy. Let ϕ_r denote the angle between the normal at point $\mathbf{r}$ and the line of sight (see Fig. 2.9). The intensity at point $\mathbf{r}$ is then the outgoing intensity in the direction of that line. Using the introduced notation convention this intensity can be denoted as $I^o_{\phi_r}$. Finally, the integrated equation of radiation can be written as [77]

$$I^i_\phi(\mathbf{p}) = I^o_{\phi_r}(\mathbf{r})\tau(\mathbf{r},\mathbf{p}) + \int_{L_{rp}} a(\mathbf{r}')\, I_b[T^m(\mathbf{r}')]\, \tau(\mathbf{r}',\mathbf{p})\, dL_{rp}(\mathbf{r}') \tag{2.28}$$

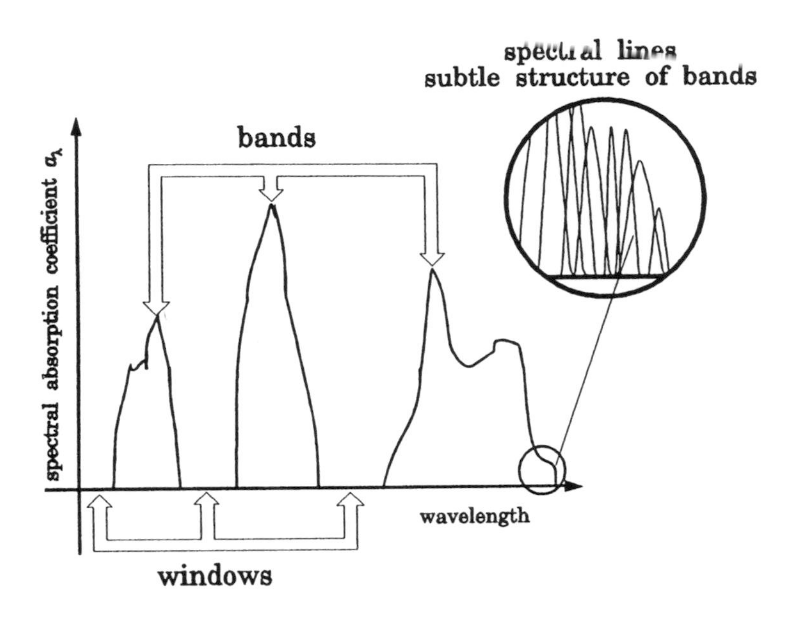

Figure 2.8: Gas radiation; sketch of the dependence of the spectral absorption coefficient on wavelength.

Functions $\tau(\mathbf{r},\mathbf{p})$, $\tau(\mathbf{r}',\mathbf{p})$ appearing in Eq.(2.28) are termed *transmissivities*. Transmissivity is defined as the fraction of energy leaving points $\mathbf{r}$ or $\mathbf{r}'$, respectively, that reaches point $\mathbf{p}$. Transmissivities are computed from the relationships

$$\begin{aligned} \tau(\mathbf{r},\mathbf{p}) &= \exp\left[-\int_{L_{rp}} a(\mathbf{r}')\, dL_{rp}(\mathbf{r}')\right] \\ \tau(\mathbf{r}',\mathbf{p}) &= \exp\left[-\int_{L_{r'p}} a(\mathbf{r}'')\, dL_{r'p}(\mathbf{r}'')\right] \end{aligned} \tag{2.29}$$

where $\mathbf{r}''$ is a point on the line of sight placed between $\mathbf{r}'$ and $\mathbf{p}$ (see Fig. 2.9)

As can be seen from Eq. (2.28) the incoming intensity at the observation point $\mathbf{p}$ is a sum of two terms:

- initial intensity attenuated by absorption within the participating medium (first right hand side term);
- emission of the medium. This radiation is also attenuated by subsequent layers of the medium passed by the ray (second right hand side term).

According to Eq. (2.25) the intensity outgoing from the surface in a direction of the line of sight can be expressed as a linear combination of radiative flux and blackbody emissive power. Taking this into account, and making use of Eq. (2.7) to

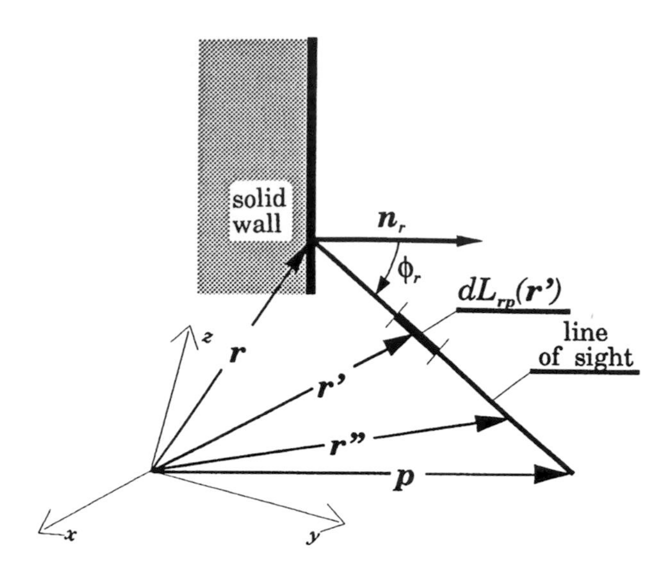

Figure 2.9: Notation used when integrating the directional equation of radiative transfer.

replace the blackbody intensity of the medium by its emissive power, Eq. (2.28) can be rewritten in the form

$$
\begin{aligned}
I^i_\phi(\mathbf{p}) &= \frac{1}{\pi}\left[e_b(\mathbf{r}) + \frac{1-\epsilon(\mathbf{r})}{\epsilon(\mathbf{r})} q^r(\mathbf{r})\right] \tau(\mathbf{r},\mathbf{p}) \\
&+ \int_{L_{rp}} a(\mathbf{r}')\, \frac{e_b[T^m(\mathbf{r}')]}{\pi}\, \tau(\mathbf{r}',\mathbf{p})\, dL_{rp}(\mathbf{r}')
\end{aligned}
\qquad (2.30)
$$

Equation (2.30) is referred to as the *integrated equation of directional radiative heat transfer*. From Eq. (2.30) intensity at an arbitrary point **p** is calculable upon performing line integration along a line of sight. To carry out this integration along an arbitrary line the distributions of the following functions should be known:

- temperature within the medium,
- absorption coefficient within the medium,
- temperature of the bounding surface,
- radiative heat flux on the bounding surface.

However, the radiative heat flux depends on the radiation incoming from all admissive directions [compare with Eq. (2.20)]. This means that the integrals along all lines of sight are interrelated. Hence, to calculate the intensity at the origin of a given line of sight, the integration over all lines ending at this point is to be carried out. The intensities at the origin of these lines are in turn related in a similar manner with intensities at other points of the enclosing surface. To take these complex interrelations into account sophisticated iteration schemes are to be employed. Therefore, directional equations of radiative transfer are fairly seldom used in engineering calculations.

2.3 Integral form of the radiative transfer equation

An alternative form of the equations governing the radiative transfer can be obtained upon substituting the integrated form of the directional transfer [Eq. (2.30)] into the heat balance on the unit surface [Eq. (2.20)]. The result reads

$$\begin{aligned} q^r(\mathbf{p}) &+ \epsilon(\mathbf{p})\, e_b[T(\mathbf{p})] \\ &= \epsilon(\mathbf{p}) \int_{2\pi} \frac{1}{\pi} \left\{ e_b[T(\mathbf{r})] + \frac{1-\epsilon(\mathbf{r})}{\epsilon(\mathbf{r})} q^r(\mathbf{r}) \right\} \tau(\mathbf{r},\mathbf{p}) \cos\phi_p \, d\Omega \\ &+ \epsilon(\mathbf{r}) \int_{2\pi} \left\{ \int_{L_{rp}} a(\mathbf{r}') \frac{e_b[T^m(\mathbf{r}')]}{\pi} \tau(\mathbf{r}',\mathbf{p})\, dL_{rp}(\mathbf{r}') \right\} \cos\phi_p \, d\Omega \end{aligned} \tag{2.31}$$

where ϕ_p denotes the angle between the line of sight and the normal at point $\mathbf{p}$.

The first integral can be converted into an integral over the surface bounding the medium. This can be accomplished upon noting that the infinitesimal solid angle can be expressed in terms of the infinitesimal area of the bounding surface

$$d\Omega = \frac{dS_\perp(\mathbf{r})}{|\mathbf{r}-\mathbf{p}|^2} = \frac{dS(\mathbf{r})\cos\phi_r}{|\mathbf{r}-\mathbf{p}|^2} \tag{2.32}$$

where $|\mathbf{r}-\mathbf{p}|$ stands for the distance between points $\mathbf{r}$ and $\mathbf{p}$ (length of the line of sight).

The second integral can be converted into a volume integral making use of obvious relationships linking the infinitesimal solid angle and the infinitesimal volume (see Fig. 2.10)

$$dL_{rp}(\mathbf{r}')d\Omega = dL_{rp}(\mathbf{r}')\frac{dS(\mathbf{r}')}{|\mathbf{r}'-\mathbf{p}|^2} = \frac{dV(\mathbf{r}')}{|\mathbf{r}'-\mathbf{p}|^2} \tag{2.33}$$

Making use of Eqs. (2.32) and (2.33) one obtains the *standard formulation of the integral equation of radiation*

$$\begin{aligned} q^r(\mathbf{p}) &+ \epsilon(\mathbf{p})e_b[T(\mathbf{p})] \\ = \; &\epsilon(\mathbf{p}) \int_S \left\{ e_b[T(\mathbf{r})] + \frac{1-\epsilon(\mathbf{r})}{\epsilon(\mathbf{r})} q^r(\mathbf{r}) \right\} \tau(\mathbf{r},\mathbf{p}) K(\mathbf{r},\mathbf{p})\, dS(\mathbf{r}) \end{aligned}$$

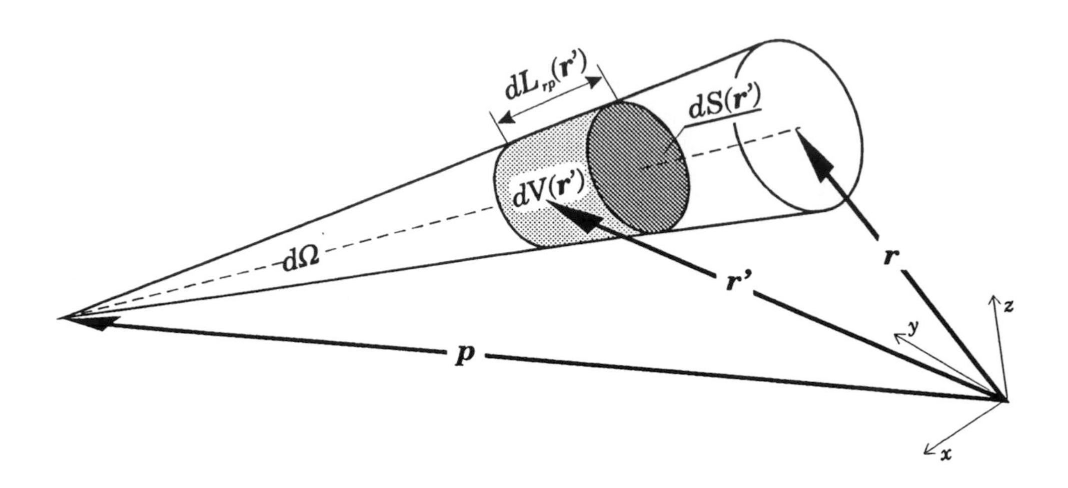

Figure 2.10: Infinitesimal solid angle and infinitesimal volume.

$$+ \quad \epsilon(\mathbf{p}) \int_V a(\mathbf{r}')\, e_b[T^m(\mathbf{r}')]\, \tau(\mathbf{r}',\mathbf{p})\; K_p(\mathbf{r}',\mathbf{p})\; dV(\mathbf{r}') \tag{2.34}$$

where the kernel functions K and K_p are defined as

$$K(\mathbf{r},\mathbf{p}) = \frac{\cos\phi_r\ \cos\phi_p}{\pi|\mathbf{r}-\mathbf{p}|^2}\beta(\mathbf{r},\mathbf{p}) \tag{2.35}$$

$$K_p(\mathbf{r}',\mathbf{p}) = \frac{\cos\phi_p}{\pi|\mathbf{r}'-\mathbf{p}|^2}\beta(\mathbf{r}',\mathbf{p}) \tag{2.36}$$

It should be noted that the kernel functions K and K_p are singular because they behave like $|\mathbf{r}-\mathbf{p}|^{-2}$ when the current point $\mathbf{r}$ tends to the observation point $\mathbf{p}$. Integrands of such order of singularity are called *strongly singular* ones.

The Boolean function β appearing in Eqs. (2.35), (2.36) takes account of the shadow zones. This function, termed the *shadow zone function*, is defined as (see Fig. 2.11)

$$\beta(\mathbf{r},\mathbf{p}) = \begin{cases} 1 & \text{if point } \mathbf{r} \text{ can be seen when looking from point } \mathbf{p} \\ 0 & \text{otherwise} \end{cases} \tag{2.37}$$

To integrals appearing in Eq. (2.34) some physical meaning can be assigned. The first integral is due to the radiation of the solid walls. The second integral is due to the radiation of the medium filling the enclosure. Both integrals contain the transmissivity, τ, which describes the attenuation of the radiation on its way through the participating medium.

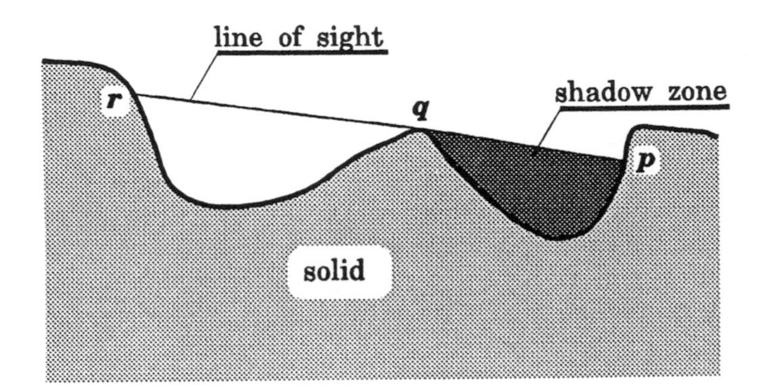

Figure 2.11: Shadow zones in heat radiation problems. When looking from point **r** boundary points placed between point **q** and **p** are in a shadow zone.

As recently shown [10] an alternative formulation of Eq. (2.31) can be derived from Eq. (2.34) substituting Eq. (2.32) into (2.33). The result reads

$$dL_{rp}(\mathbf{r}')\frac{dS(\mathbf{r})\cos\phi_r}{|\mathbf{r}-\mathbf{p}|^2} = \frac{dV(\mathbf{r}')}{|\mathbf{r}'-\mathbf{p}|^2} \tag{2.38}$$

from which it follows that for the arbitrary regular function $f(\mathbf{r},\mathbf{p})$ a relationship linking the volume and surface integral holds

$$\int_V \frac{f(\mathbf{r}',\mathbf{p})}{|\mathbf{r}'-\mathbf{p}|^2}\, dV(\mathbf{r}') = \int_S \left\{ \int_{L_{rp}} f(\mathbf{r}',\mathbf{p})\, dL_{rp}(\mathbf{r}') \right\} \frac{\cos\phi_r}{|\mathbf{r}-\mathbf{p}|^2}\, dS(\mathbf{r}) \tag{2.39}$$

In Eq. (2.39) point **r** lies on the bounding surface S whereas point $\mathbf{r}'$ is within the domain. Point **p** can be placed either on the bounding surface or within the domain. Equation (2.39) shows that a volume integral can be converted into a surface integral from a line integral.

With this in mind the equation of radiative transfer (2.34) can be written as

$$\begin{aligned} & q^r(\mathbf{p}) + \epsilon(\mathbf{p})e_b[T(\mathbf{p})] \\ = \; & \epsilon(\mathbf{p}) \int_S \left\{ e_b[T(\mathbf{r})] + \frac{1-\epsilon(\mathbf{r})}{\epsilon(\mathbf{r})} q^r(\mathbf{r}) \right\} \tau(\mathbf{r},\mathbf{p}) K(\mathbf{r},\mathbf{p})\, dS(\mathbf{r}) \\ + \; & \epsilon(\mathbf{p}) \int_S \left\{ \int_{L_{rp}} a(\mathbf{r}')\, e_b[T^m(\mathbf{r}')]\, \tau(\mathbf{r}',\mathbf{p})\, dL_{rp}(\mathbf{r}') \right\} K(\mathbf{r},\mathbf{p})\, dS(\mathbf{r}) \end{aligned} \tag{2.40}$$

The physical meaning of both integral terms on the right hand side of Eq. (2.40) is the same as those occurring in Eq. (2.34). Formulations (2.34) and (2.40) are completely equivalent. However, one should notice that no volume integral is present in Eq. (2.40). Instead there is a double integral over the bounding surface and along the line of sight. It will be shown in subsequent chapters that the integration along the line of sight can be, in most cases, performed analytically. Thus, the volume integral occurring in Eq. (2.34) has been converted into a surface integral. This is equivalent to the reduction of the dimensionality of the problem and results in a significant economy of the computer time needed for the discretization of the integral equation. This feature of Eq. (2.40) is similar to the BEM technique where the reduction of the dimensionality is also carried out (see Section 1.2). Equation (2.40) will be referred to as the *BEM formulation of the integral equation of heat radiation.*

Provided material properties (ϵ and a) are known, integral equations (2.34) and (2.40) describe the interrelation of three functions:

- emissive power (temperature) $e_b[T(\mathbf{r})]$ of the walls of the enclosure,
- emissive power (temperature) $e_b[T^m(\mathbf{r}')]$ of the medium filling the enclosure,
- radiative heat flux on the walls of the enclosure $q^r(\mathbf{r})$.

Equations to compute the radiative heat source can be obtained from the radiative energy balance on the infinitesimal volume (2.27). Substituting the integrated form of the directional equation of transfer (2.30) into (2.27) and following the same way of reasoning as already described, one ends up with two formulations as follows

standard formulation

$$\begin{aligned} q_V^r(\mathbf{p}) &+ 4a(\mathbf{p})e_b[T^m(\mathbf{p})] \\ &= a(\mathbf{p})\int_S \left\{ e_b[T(\mathbf{r})] + \frac{1-\epsilon(\mathbf{r})}{\epsilon(\mathbf{r})} q^r(\mathbf{r}) \right\} \tau(\mathbf{r},\mathbf{p}) K_r(\mathbf{r},\mathbf{p})\, dS(\mathbf{r}) \\ &+ a(\mathbf{p})\int_V a(\mathbf{r}')\, e_b[T^m(\mathbf{r}')]\, \tau(\mathbf{r}',\mathbf{p})\, K_0(\mathbf{r}',\mathbf{p})\, dV(\mathbf{r}') \end{aligned} \tag{2.41}$$

BEM formulation

$$\begin{aligned} & q_V^r(\mathbf{p}) + 4a(\mathbf{p})e_b[T^m(\mathbf{p})] \\ = \; & a(\mathbf{p})\int_S \left\{ e_b[T(\mathbf{r})] + \frac{1-\epsilon(\mathbf{r})}{\epsilon(\mathbf{r})} q^r(\mathbf{r}) \right\} \tau(\mathbf{r},\mathbf{p}) K_r(\mathbf{r},\mathbf{p})\, dS(\mathbf{r}) \\ + \; & a(\mathbf{p})\int_S \left\{ \int_{L_{rp}} a(\mathbf{r}')\, e_b[T^m(\mathbf{r}')]\, \tau(\mathbf{r}',\mathbf{p})\, dL_{rp}(\mathbf{r}') \right\} K_r(\mathbf{r},\mathbf{p})\, dS(\mathbf{r}) \end{aligned} \tag{2.42}$$

where the kernel functions K_r and K_0 are defined as

$$K_r(\mathbf{r},\mathbf{p}) = \frac{\cos\phi_r}{\pi|\mathbf{r}'-\mathbf{p}|^2}\beta(\mathbf{r},\mathbf{p}) \tag{2.43}$$

$$K_0(\mathbf{r}',\mathbf{p}) = \frac{1}{\pi|\mathbf{r}'-\mathbf{p}|^2}\beta(\mathbf{r}',\mathbf{p}) \tag{2.44}$$

Physical interpretation of the integrals occurring in Eqs. (2.41) and (2.42) is identical as in the case of Eqs. (2.34) and (2.40). It is worth noting that the volume integral present in formulation (2.41) has, in Eq. (2.42), been converted into a surface integral.

It should also be stressed that the radiative heat source does not appear under the integral sign. Thus, Eqs. (2.41) and (2.42) are not integral equations with respect to the radiative heat source q_V^r. Once temperatures of both the medium and the bounding surface, as well as the radiative heat fluxes are known, the radiative heat source can be computed explicitly by carrying out appropriate integrations.

The kernel functions K, K_p, K_r and K_0 occurring in the equations of radiative transfer (2.34), (2.40) to (2.42) exhibit singular behaviour when the distance between the observation point $\mathbf{p}$ and current point $\mathbf{r}$ tends to zero. The order of singularity of these kernels is $|\mathbf{p}-\mathbf{r}|^{-2}$.

2.4 Specific cases

2.4.1 Transparent medium

Radiation on its way through a transparent medium is neither attenuated nor emitted and the radiative heat source function vanishes throughout the domain. Moreover, in the presence of transparent media, radiative transfer takes place solely between surfaces bounding the medium. Equations corresponding to this situation can be readily obtained from Eqs. (2.34) or (2.40) upon putting the absorption coefficient a equal to zero (note that this implies that the transmissivity is equal to 1). The result reads

$$\begin{aligned} q^r(\mathbf{p}) &+ \epsilon(\mathbf{p})e_b[T(\mathbf{p})] \\ &= \epsilon(\mathbf{p}) \int_S \left\{ e_b[T(\mathbf{r})] + \frac{1-\epsilon(\mathbf{r})}{\epsilon(\mathbf{r})} q^r(\mathbf{r}) \right\} K(\mathbf{r},\mathbf{p})\, dS(\mathbf{r}) \end{aligned} \tag{2.45}$$

Equation (2.45) links the emissive powers (temperatures) of the mutually irradiating surfaces with the radiative heat flux gained by these surfaces.

2.4.2 Isothermal, participating medium of constant absorption coefficient

When the temperature and composition of the medium can be regarded as constant, absorption coefficient and blackbody emissive powers are position independent. For this specific case the integrals along the line of sight entering equations of transfer can be readily calculated. The result of the analytical integration is

$$\tau(\mathbf{r},\mathbf{p}) = \exp\left[-\int_{L_{rp}} a(\mathbf{r}')\, dL_{rp}(\mathbf{r}')\right] = \exp\left(-a|\mathbf{r}-\mathbf{p}|\right) \tag{2.46}$$

$$\int_{L_{rp}} a(\mathbf{r}')\, e_b[T^m(\mathbf{r}')]\, \tau(\mathbf{r}',\mathbf{p})\, dL_{rp}(\mathbf{r}') = e_b(T^m)\,[1 - \exp(-a|\mathbf{r}-\mathbf{p}|)] \tag{2.47}$$

By virtue of Eqs. (2.46) and (2.47) equations of transfer (2.40) and (2.42) can be rewritten in a form

$$\begin{aligned} q^r(\mathbf{p}) \;&+\; \epsilon(\mathbf{p}) e_b[T(\mathbf{p})] \\ &= \epsilon(\mathbf{p}) \int_S \left\{ e_b[T(\mathbf{r})] + \frac{1-\epsilon(\mathbf{r})}{\epsilon(\mathbf{r})} q^r(\mathbf{r}) \right\} K(\mathbf{r},\mathbf{p}) \exp(-a|\mathbf{r}-\mathbf{p}|)\, dS(\mathbf{r}) \\ &+ \epsilon(\mathbf{p})\, e_b(T^m) \int_S K(\mathbf{r},\mathbf{p})\, [1 - \exp(-a|\mathbf{r}-\mathbf{p}|)]\; dS(\mathbf{r}) \end{aligned} \tag{2.48}$$

$$\begin{aligned} q_V^r(\mathbf{p}) \;&+\; 4a\, e_b(T^m) \\ &= a \int_S \left\{ e_b[T(\mathbf{r})] + \frac{1-\epsilon(\mathbf{r})}{\epsilon(\mathbf{r})} q^r(\mathbf{r}) \right\} K_r(\mathbf{r},\mathbf{p}) \exp(-a|\mathbf{r}-\mathbf{p}|)\, dS(\mathbf{r}) \\ &+ a\, e_b(T^m) \int_S K_r(\mathbf{r},\mathbf{p})\, [1 - \exp(-a|\mathbf{r}-\mathbf{p}|)]\; dS(\mathbf{r}) \end{aligned} \tag{2.49}$$

Similar equations can be obtained when starting from the standard formulation, i.e. Eqs. (2.34) and (2.41)

$$\begin{aligned} q^r(\mathbf{p}) \;&+\; \epsilon(\mathbf{p}) e_b[T(\mathbf{p})] \\ &= \epsilon(\mathbf{p}) \int_S \left\{ e_b[T(\mathbf{r})] + \frac{1-\epsilon(\mathbf{r})}{\epsilon(\mathbf{r})} q^r(\mathbf{r}) \right\} K(\mathbf{r},\mathbf{p}) \exp(-a|\mathbf{r}-\mathbf{p}|)\, dS(\mathbf{r}) \\ &+ \epsilon(\mathbf{p})\, a\, e_b(T^m) \int_V K_p(\mathbf{r}',\mathbf{p}) \exp(-a|\mathbf{r}'-\mathbf{p}|)\, dV(\mathbf{r}') \end{aligned} \tag{2.50}$$

$$\begin{aligned} q_V^r(\mathbf{p}) \;&+\; 4a\, e_b(T^m) \\ &= a \int_S \left\{ e_b[T(\mathbf{r})] + \frac{1-\epsilon(\mathbf{r})}{\epsilon(\mathbf{r})} q^r(\mathbf{r}) \right\} K_r(\mathbf{r},\mathbf{p}) \exp(-a|\mathbf{r}-\mathbf{p}|)\, dS(\mathbf{r}) \\ &+ a^2\, e_b(T^m) \int_V K_0(\mathbf{r}',\mathbf{p}) \exp(-a|\mathbf{r}'-\mathbf{p}|)\, dV(\mathbf{r}') \end{aligned} \tag{2.51}$$

2.4.3 Nongray, nonisothermal medium; band approximation

The spectral absorption coefficient a_λ of a nongray medium depends, by definition, on the wavelength. In the presence of such media, spectral radiative heat flux q_λ^r and radiative heat source $q_{\lambda V}^r$ are also wavelength dependent. Usually the total counterparts of these functions, i.e. q^r and q_V^r are of interest. Due to Eq. (2.3) these functions are interrelated by the relationships

$$q^r = \int_0^\infty q_\lambda^r \, d\lambda \tag{2.52}$$

$$q_V^r = \int_0^\infty q_{\lambda V}^r \, d\lambda \tag{2.53}$$

Spectral quantities entering Eqs. (2.52) and (2.53) can be obtained upon solving integral equation sets written for a certain wavelength. Standard formulations [Eqs. (2.34), (2.41)] or their BEM counterpart [Eqs. (2.40), (2.42)] can be used. Recalling the notation convention for total and spectral quantities, the (spectral) equations are obtained upon appending the subscript λ to following quantities:

- radiative heat flux; q^r,
- radiative heat source; q_V^r,
- blackbody emissive power; e_b,
- emissivity; ϵ,
- absorption coefficient; a.

Total quantities are then calculated upon replacing the integrals appearing in Eqs. (2.52) and (2.53) by appropriate numerical quadratures. This approach can lead to significant errors when the medium under consideration is a gas. Due to quantum effects the absorption coefficient of gases depends, in a very irregular manner, on the wavelength (see Fig. 2.8). Unless a very fine subdivision of the spectrum is used, numerical integration over the entire spectrum does not yield accurate results.

As already stated, the spectral absorption coefficient of gases vanishes within certain intervals of the spectrum (windows) while assuming nonzero values within other intervals termed bands. One of the simplest methods of taking into account this spectral behaviour of radiation in gases, is to assume that the absorption coefficient within bands remains constant. For each band the mean (equivalent) value of the absorption coefficient is used in the computations. The value of the absorption coefficient assigned to windows is obviously zero.

Let the entire spectrum be divided into a finite number of intervals:

$$\Delta\lambda^u = (\lambda^u, \lambda^{u+1}); \quad u = 1, \ldots, U \tag{2.54}$$

The emissivity and absorption coefficient (assumed not temperature dependent) associated with the uth band are denoted by ϵ^u and a^u, respectively. New unknowns defined as

$$q^{ru}(\mathbf{r}) = \int_{\lambda^u}^{\lambda^{u+1}} q_\lambda^r(\lambda', \mathbf{r})\, d\lambda' \tag{2.55}$$

$$q_V^{ru}(\mathbf{r}) = \int_{\lambda^u}^{\lambda^{u+1}} q_{\lambda V}^r(\lambda', \mathbf{r})\, d\lambda' \tag{2.56}$$

$$e_b^u(\mathbf{r}) = \int_{\lambda^u}^{\lambda^{u+1}} e_{b\lambda}(\lambda', \mathbf{r})\, d\lambda' \tag{2.57}$$

are introduced. Provided the temperature is known, the third unknown can be computed from Wiebelt's approximation [see Eq. (2.13)]. The resulting integral equation corresponding to the uth band can be obtained upon appending the superscript u to appropriate functions, i.e. putting

$$a := a^u \quad \epsilon := \epsilon^u \quad e_b := e_b^u \quad q^r := q^{ru} \quad q_V^r := q_V^{ru} \quad \tau := \tau^u \tag{2.58}$$

where

$$\tau^u(\mathbf{r},\mathbf{p}) = \exp\left[-\int_{L_{rp}} a^u(\mathbf{r}')\, dL_{rp}(\mathbf{r}')\right] \tag{2.59}$$

Equations corresponding to bands are obtained by performing substitutions defined by Eq. (2.58), in Eqs. (2.34) and (2.41) (standard formulation) or alternatively in Eqs. (2.40) and (2.42) (BEM formulation). The latter have an appearance

$$\begin{aligned} & q^{ru}(\mathbf{p}) + \epsilon^u(\mathbf{p}) e_b^u[T(\mathbf{p})] \\ = \; & \epsilon^u(\mathbf{p}) \int_S \left\{ e_b^u[T(\mathbf{r})] + \frac{1-\epsilon^u(\mathbf{r})}{\epsilon^u(\mathbf{r})} q^{ru}(\mathbf{r}) \right\} \tau^u(\mathbf{r},\mathbf{p}) K(\mathbf{r},\mathbf{p})\, dS(\mathbf{r}) \\ + \; & \epsilon^u(\mathbf{p}) \int_S \left\{ \int_{L_{rp}} a^u(\mathbf{r}')\, e_b^u[T^m(\mathbf{r}')]\, \tau^u(\mathbf{r}',\mathbf{p})\, dL_{rp}(\mathbf{r}') \right\} K(\mathbf{r},\mathbf{p})\, dS(\mathbf{r}) \end{aligned} \tag{2.60}$$

$$\begin{aligned} & q_V^{ru}(\mathbf{p}) + 4a^u(\mathbf{p}) e_b^u[T^m(\mathbf{p})] \\ = \; & a^u(\mathbf{p}) \int_S \left\{ e_b^u[T(\mathbf{r})] + \frac{1-\epsilon^u(\mathbf{r})}{\epsilon^u(\mathbf{r})} q^{ru}(\mathbf{r}) \right\} \tau^u(\mathbf{r},\mathbf{p}) K_r(\mathbf{r},\mathbf{p})\, dS(\mathbf{r}) \\ + \; & a^u(\mathbf{p}) \int_S \left\{ \int_{L_{rp}} a^u(\mathbf{r}')\, e_b^u[T^m(\mathbf{r}')]\, \tau^u(\mathbf{r}',\mathbf{p})\, dL_{rp}(\mathbf{r}') \right\} K_r(\mathbf{r},\mathbf{p})\, dS(\mathbf{r}) \end{aligned} \tag{2.61}$$

To obtain values corresponding to windows, substitutions (2.58) are to be carried out in Eq. (2.45). The result reads

$$\begin{aligned} q^{ru}(\mathbf{p}) \; + \; & \epsilon^u(\mathbf{p}) e_b^u[T(\mathbf{p})] \\ = \; & \epsilon^u(\mathbf{p}) \int_S \left\{ e_b^u[T(\mathbf{r})] + \frac{1-\epsilon^u(\mathbf{r})}{\epsilon^u(\mathbf{r})} q^{ru}(\mathbf{r}) \right\} K(\mathbf{r},\mathbf{p})\, dS(\mathbf{r}) \end{aligned} \tag{2.62}$$

Knowing the heat associated with subsequent spectral intervals, the total heat can be computed as a sum of interval contributions:

$$q^r = \sum_{u=1}^{U} q^{ru} \tag{2.63}$$

$$q_V^r = \sum_{u=1}^{U} q_V^{ru} \tag{2.64}$$

As the radiative heat source within windows vanishes, the summation index u in Eq. (2.64) runs only over bands.

2.4.4 Nongray participating medium; weighted sum of gray gases approximation

To cope with the complex nature of gas radiation many approximate methods have been developed. One of the most popular techniques used in engineering calculations is the *weighted sum of gray gas* approach. The method does not have a sound physical background, but due to the ease of implementation it is very popular among modellers. The main idea [40] is to represent the radiation of gas by a sum of terms, each being a product of a weighting factor and an equivalent gray gas emission. The weighting factors are assumed temperature dependent while the associated equivalent gray gas absorption coefficients are constant. The weighting factors are expressed in terms of polynomials of the emitter temperature. The coefficients of the polynomials, as well as the gray gas absorption coefficients, are determined by fitting total gas emissivity data [79]. The latter are obtained either from spectral data by quantum mechanics computations or from measurements. To take into account the radiation transferred within the windows the absorption coefficient of the first term of the sum of gray gases is taken as equal to zero. Depending on the number of terms in this sum the models are named *one clear n gray gas* for n nonzero absorption coefficients considered.

The formula to determine the coefficients of the model is

$$\epsilon^m = \sum_{i=0}^{n} v_i(T^m)\,[1 - \exp(-a_i L_{rp})] \tag{2.65}$$

where v_i stands for the weighting factors , a_i are the equivalent gray gas absorption coefficients, L_{rp} is the length of the line of sight, ϵ^m is the emissivity of the gas. This property is defined analogously to the surface emissivity. Consider an isothermal volume of gas of temperature T^m. The radiant energy leaving this volume is less than the emission of a blackbody having the same temperature. The emissivity of a gas volume is defined as a quotient of these two radiant energies.

If the number of terms in Eq. (2.65) is very large v_i can be interpreted as an energy fraction of the blackbody emission in all these spectral intervals where the absorption coefficient is about a_i. However, more properly a_i and v_i should be considered as numbers which make the approximation formula (2.65) fit a given function. To make the approximation consistent, weighting coefficients should satisfy several limitations. They will now be discussed briefly.

Thick layers of participating gas absorb entire incident radiation, thus a large gas volume behaves as a blackbody. Similarly, emission of an isothermal gas volume approaches a blackbody limit when the length of the line of sight tends to infinity. This behaviour is taken into account by imposing constraints on the sum of the weighting functions

$$\sum_{i=0}^{n} v_i(T^m) = 1 \tag{2.66}$$

Each gray gas should give a positive contribution to the total gas emission thus the weighting functions should be positive

$$v_i(T^m) > 0 \tag{2.67}$$

To take into account the presence of spectral windows , where the gas is transparent to radiation, the absorption coefficient associated with the zeroth gray gas is taken as zero

$$a_0 = 0 \tag{2.68}$$

The equations of heat radiation are solved separately using subsequent gray gases absorption coefficients. The total radiation is calculated as a weighted sum of contributions coming from subsequent gray gases. The details of this technique can be found in the literature [41].

Modelling real gas radiation by a sum of gray gases misrepresents some physical realities. Better physical interpretation of the approximating formula can be achieved by using a hybrid *band-weighted sum of gray gases model* approach. The first step of this technique is the same as in the band approximation model, i.e. the subdivision of the entire spectrum into a finite number of intervals. Within some intervals (windows) gas is assumed transparent to radiation. Gas emission in the remaining intervals (bands) is approximated by *one gray gas* term with the weighting function depending on temperature, composition and pressure [5, 83, 84].

2.4.5 2D problems

For a situation when all bodies taking part in the radiative heat exchange are elongated in one direction, the original 3D problem can be treated as a 2D one. Integral equations derived in this chapter remain valid also for 2D analysis. To use these equations in 2D analysis the kernel functions $K(\mathbf{r},\mathbf{p})$, $K_p(\mathbf{r},\mathbf{p})$, $K_r(\mathbf{r},\mathbf{p})$, $K_0(\mathbf{r},\mathbf{p})$ appearing in these equations should be redefined as follows:

$$K(\mathbf{r},\mathbf{p}) = \frac{\cos\phi_r \cos\phi_p}{2|\mathbf{r}-\mathbf{p}|} \tag{2.69}$$

$$K_r(\mathbf{r},\mathbf{p}) = \frac{\cos\phi_r}{2|\mathbf{r}-\mathbf{p}|} \tag{2.70}$$

$$K_p(\mathbf{r},\mathbf{p}) = \frac{\cos\phi_p}{2|\mathbf{r}-\mathbf{p}|} \tag{2.71}$$

$$K_0(\mathbf{r},\mathbf{p}) = \frac{1}{2|\mathbf{r}-\mathbf{p}|} \tag{2.72}$$

Chapter 3

Weighted residuals method

3.1 Analytical versus numerical methods

Mathematical models of physical phenomena encountered in science and engineering are formulated as sets of differential or/and integral equations. The methods of solving these equations fall into two groups; analytical and numerical. Efficient usage of analytical methods is restricted to linear problems in domains of simple shapes. Irregular boundaries of the majority of practical problems preclude any analytical solution. Moreover, in order to assure sufficient accuracy, nonlinearity present in the boundary conditions, material behaviour, etc., should often be taken into account. Complex geometry and nonlinearity can be handled using numerical techniques with relative ease. Moreover, some numerical techniques enable one to build general purpose computer codes capable of dealing with different physical situations, bodies of arbitrary shape, any type of nonlinearity. Many commercial codes of this generality are available. Owing to the flexibility and ease of usage of numerical methods, they became the only feasible means of obtaining adequately precise and detailed results. Therefore the bulk of engineering problems are nowadays solved using codes based on these techniques.

Numerical techniques transform the equations of mathematical models into a set of algebraic equations which introduces additional error. Analytical solutions, when obtainable, are free from this error, therefore they serve as a source of benchmark solutions used to test accuracy and correctness of results obtained via numerical techniques.

The most frequently used numerical methods are, as mentioned in Section 1.2, Finite Differences, Finite Element and Boundary Element Methods. A common feature of these methods is that they can be interpreted as specific cases of the *weighted residuals method– WRM* [23, 87]. Within the present chapter the general philosophy of a specific version of the WRM will be discussed. An exhaustive description of WRM can be found elsewhere [33].

WRM is a general numerical technique capable of solving ordinary and partial differential equations as well as integral equations. Because only integral equations are dealt with in the present book, the discussion of WRM will be restricted solely to integral equations. The equations can be formulated in a space of arbitrary dimension. In order to simplify the notation 1D integral equations will be considered.

3.2 Approximation of function

Consider an integral equation having a form

$$Df(y) + C\int_c^d f(x)k(x,y)\,dx = g(y) \tag{3.1}$$

where:

$f(x)$– sought for function,

$g(y)$– known function,

$k(x,y)$– known kernel function,

(c,d)– integration interval,

C, D– known scalars.

Integral equations are formulated in the space of functions, hence the number of degrees of freedom to be found is infinite. A characteristic feature of all numerical techniques is that this number is made finite. In WRM this is accomplished upon approximating the unknown function by a linear combination of a finite number of known functions. The interpolation formula has a general form

$$f(x) \approx \sum_{m=1}^{M} \alpha_m \Phi_m(x) \tag{3.2}$$

where $\Phi_m(x)$ denotes known functions and α_m; $m = 1, 2, \ldots M$ are the sought after scalars.

Substitution of Eq. (3.2) into (3.1) yields

$$\sum_{m=1}^{M} \alpha_m h_m(y) - g(y) \approx 0 \tag{3.3}$$

where

$$h_m(y) = D\Phi_m(y) + C\int_c^d k(x,y)\Phi_m(x)\,dx \tag{3.4}$$

3.3 Weighting the residuals

If formula (3.2) was the solution of Eq. (3.1) then Eq. (3.3) would be satisfied for arbitrary y. As Eq. (3.2) is only an approximation of the solution, Eq. (3.3) cannot be satisfied for each y. The measure of the error introduced by the approximation of the unknown function by the interpolation formula (3.2) is referred to as the *residuum* and defined as

$$r(y) = g(y) - \sum_{m=1}^{M} \alpha_m h_m(y) \tag{3.5}$$

As the residuum does not vanish for any y, to obtain a reasonable approximation of the solution one can 'at least' minimize this error. There are many possibilities of performing such a minimization. The Weighted Residuals Method requires that the integral of a product of the residuum and appropriately chosen function $w(y)$, called *weighting function*, vanishes, i.e.

$$\int_c^d w(y) r(y)\, dy = \int_c^d w(y) \left[g(y) - \sum_{m=1}^{M} \alpha_m h_m(y) \right] dy = 0 \tag{3.6}$$

After carrying out appropriate integrations Eq. (3.6) gives a linear relationship linking coefficients α_m. One needs M such equations to determine all these scalars. Appropriate relationships are generated upon repeating the weighting procedure (3.6) M times, taking as weighting functions subsequent elements of an appropriately chosen sequence of functions $w_j(y)$. The coefficients α_m of the approximating function (3.2) are then computed upon solving a linear set of equations

$$\mathbf{W}\boldsymbol{\alpha} = \mathbf{g} \tag{3.7}$$

where elements of the matrices arising in (3.7) are defined as

$$\{\mathbf{W}\}_{jm} = \int_c^d w_j(y) h_m(y)\, dy \tag{3.8}$$

$$\{\boldsymbol{\alpha}\}_m = \alpha_m \tag{3.9}$$

$$\{\mathbf{g}\}_j = \int_c^d w_j(y) g(y)\, dy \tag{3.10}$$

3.4 Choice of interpolating and weighting functions

To form the WRM matrix $\mathbf{W}$ two sequences of functions are to be chosen. These are:

- interpolating functions Φ_m,
- weighting functions w_j.

Extensive analysis of the question of a proper choice of these two sets of functions is addressed in mathematical literature. To follow this reasoning a knowledge of sophisticated mathematical tools is essential. No attempt is therefore made to repeat this analysis here. Instead, some practical hints will be given using intuition rather than a sound mathematical theory.

To assure that the resulting WRM matrix **W** is nonsingular, both sequences should consist of *linearly independent functions.* A set of function Φ_m; $m = 1, 2, \ldots M$ is said to be linearly independent, when no function belonging to this set can be expressed as a finite sum of products of arbitrary scalars α_m and other elements of the set, i.e. when the equation

$$\sum_{m=1}^{M} \alpha_m \Phi_m(x) = 0 \tag{3.11}$$

is satisfied only when all α_m simultaneously vanish.

It is a natural expectation that the increase of the number of degrees of freedom M should lead to better final results. This is a somewhat simplified definition of the *convergence* of a numerical method. To possess this important feature the set of interpolating functions Φ_m should be *complete.* When a sequence of functions having this property is used, a function belonging to a class of interest can be approximated with any prescribed accuracy. If Φ_m constitutes a complete set of functions formula (3.2) is inherently capable of representing the exact solution, provided enough terms are used.

Completeness and linear independence of functions are the primary criteria of choosing the weighting and interpolating functions. Additional criteria are connected with the numerical efficiency of the approach. A proper choice of both function sequences Φ_m and w_j leads to substantial computer time saving. The functions should be taken in such a way, that the integrations present in Eqs. (3.4), (3.8), and (3.10) can be carried out cheaply. As in practical implementations of WRM the integration is carried out numerically, this requirement is equivalent to a simple form of functions arising in both sequences.

Computer software approximates all transcendental functions by polynomials. Thus, the idea of taking polynomials as interpolating functions naturally arises. However, an unpleasant feature of interpolation using higher order polynomials is the ill conditioning of the resulting matrices and large oscillations of the approximating polynomial. To circumvent this difficulty, locally based, lower order polynomials known as *shape functions* are often used. A characteristic feature of these functions is that they vanish outside a finite interval. To use shape functions as interpolating functions the entire interpolation interval is subdivided into a number of subintervals (elements). The order of the shape functions associated with this element depends on the number of nodes placed on the element. Consider an element ΔS_e with x_i $i = 1, 2, \ldots, I^e$ nodal points placed on it. The shape functions Φ_i^e associated with this element are polynomials of order $I^e - 1$ having within ΔS_e a property

$$\Phi_i^e(x_j) = \delta_{ij} \tag{3.12}$$

where δ_{ij} stands for Kronecker's symbol defined as

$$\delta_{ij} = \begin{cases} 1 & \text{if } i = j \\ 0 & \text{otherwise} \end{cases} \tag{3.13}$$

and vanishing outside ΔS_e. The latter property causes that integration of integrands being a product of an arbitrary function $F(x)$ and a shape function is always limited to a single element within which the shape function does not vanish, i.e.

$$\int_c^d \Phi_i^e(x)F(x)dx = \int_{\Delta S_e} \Phi_i^e(x)F(x)dx \tag{3.14}$$

It should also be pointed out that with the shape functions defined by Eq. (3.12) the sought for coefficients α_m of the interpolation function (3.2) are just the nodal values of the unknown function. Thus, some physical meaning can be assigned to the unknowns, which is another advantage of employing the shape function concept.

Shape functions are usually expressed in terms of local coordinates ξ. Local coordinates are defined in such a way that the interval $\xi \in [-1, 1]$ covers the entire element ΔS_e. Below are some examples of shape functions (see Fig. 3.1).

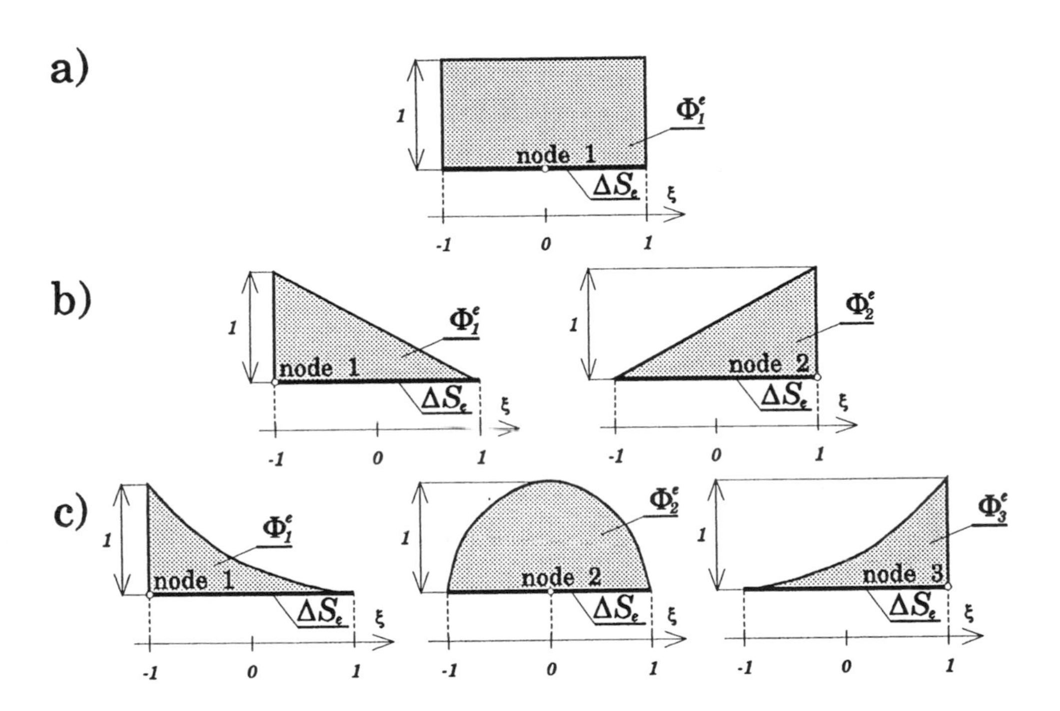

Figure 3.1: Shape functions: *a*) constant, *b*) linear, *c*) quadratic.

If only one nodal point is placed on the element ($I^e = 1$), this node is usually centered. Thus, the shape function is in this case associated with a point of local coordinate $\xi = 0$. The shape function is then defined as

$$\Phi_1^e(\xi) = \begin{cases} 1 & \text{if } \xi \in \Delta S_e \\ 0 & \text{otherwise} \end{cases} \tag{3.15}$$

Linear approximation of the function is accomplished upon placing two nodes ($I^e = 2$) at both ends of the element. Thus, the shape functions are associated with nodes of coordinates $\xi = +1$ and $\xi = -1$.

$$\begin{aligned} \Phi_1^e(\xi) &= (1-\xi)/2 \quad \text{nodal point } \xi = -1 \\ \Phi_2^e(\xi) &= (1+\xi)/2 \quad \text{nodal point } \xi = +1 \end{aligned} \tag{3.16}$$

Quadratic approximation of the function is accomplished upon placing three nodes ($I^e = 3$): one in the center and two at both ends of the element. Thus, the shape functions are associated with nodes of coordinates $\xi = +1$ $\xi = 0$ and $\xi = -1$.

$$\begin{aligned} \Phi_1^e(\xi) &= (\xi-1)\xi/2 && \text{nodal point } \xi = -1 \\ \Phi_2^e(\xi) &= (1+\xi)(1-\xi) && \text{nodal point } \xi = 0 \\ \Phi_3^e(\xi) &= (\xi+1)\xi/2 && \text{nodal point } \xi = +1 \end{aligned} \tag{3.17}$$

Note that except for Eq. (3.15) shape functions ensure continuity of the interpolation polynomials between adjacent elements.

The version of the WRM using the same set of functions to approximate both the unknown function and to weight the residuals ($\Phi_m = w_m$) is known as the *Galerkin method.*

Simple formulae for the WRM matrices are obtained upon taking, as the weighting functions, *Dirac's distributions* δ acting at subsequent interpolation nodes y_j. The filtration property of the δ function is defined as

$$\int_{\Delta S_e} f(y)\delta(y_j - y)\,dy = \begin{cases} f(y_j) & \text{if } y_j \in \Delta S_e \\ 0.5\,f(y_j) & \text{if } y_j \text{ is a boundary point of } \Delta S_e \\ 0 & \text{if } y_j \text{ lies outside } \Delta S_e \end{cases} \tag{3.18}$$

Due to property (3.18) integration in Eq. (3.6) degenerates to a substitution of the nodal point coordinate into the integrand. Thus, Eq. (3.6) acquires a form

$$\int_c^d \delta(y_j - y) r(y)\,dy = g(y_j) - \sum_{m=1}^{M} \alpha_m h_m(y_j) = 0 \tag{3.19}$$

The version of the Weighted Residuals Methods where Dirac's distributions are used as the weighting functions is known as the *collocation method.* The points being the arguments of Dirac's distributions are called *collocation points.*

3.5 Example

The application of the Weighted Residuals Method will be shown on a simple example. Sought for is the solution $f(x)$ of a linear integral equation

$$f(y) + \int_0^1 (y - x) f(x) dx = \frac{4y-3}{12} + y^2 \tag{3.20}$$

The exact solution of Eq. (3.20) reads

$$f(x) = x^2 \tag{3.21}$$

Four variants of WRM will be discussed:

i. collocation with shape functions constant within elements,

ii. Galerkin with constant shape functions,

iii. collocation with linear shape functions ,

iv. Galerkin with linear shape functions .

The integration interval [0–1] is subdivided into two equal subintervals (elements) [0–0.5] and [0.5–1]. Consider cases *i* and *ii*. Let $f_{\frac{1}{4}}$ and $f_{\frac{3}{4}}$ denote the unknown values of the sought for function at nodes $x = 0.25$ and $x = 0.75$, respectively. The unknown function can be approximated by an expression

$$f(x) \approx f_{\frac{1}{4}}\Phi_1^1(x) + f_{\frac{3}{4}}\Phi_1^2(x) \tag{3.22}$$

where the shape functions are defined as (see Fig. 3.2)

$$\Phi_1^1(x) = \begin{cases} 1 & \text{if } 0 \le x \le 0.5 \\ 0 & \text{if } 0.5 < x \le 1 \end{cases} \tag{3.23}$$

$$\Phi_1^2(x) = \begin{cases} 0 & \text{if } 0 \le x \le 0.5 \\ 1 & \text{if } 0.5 < x \le 1 \end{cases} \tag{3.24}$$

Equation (3.22) is substituted into (3.20) and the global coordinates x, y replaced by their local counterpart $\xi \in [-1, 1]$. The relationship between local and global coordinates is defined by a linear transformation

$$x, y = \begin{cases} \dfrac{1+\xi}{4} & \text{if } 0 \le x, y \le 0.5 \\[2ex] \dfrac{3+\xi}{4} & \text{if } 0.5 < x, y \le 1 \end{cases} \tag{3.25}$$

After performing the integration two expressions for the residuum are obtained

$$\begin{aligned} r[y(\xi)] &= f_{\frac{1}{4}}\frac{\xi+8}{8} + f_{\frac{3}{4}}\frac{\xi-2}{8} - \frac{3\xi^2+10\xi-5}{48}; \\ & \qquad y \in [0-0.5],\ \xi \in [-1, 1] \end{aligned} \tag{3.26}$$

$$\begin{aligned} r[y(\xi)] &= f_{\frac{1}{4}}\frac{\xi+2}{8} + f_{\frac{3}{4}}\frac{\xi+8}{8} - \frac{3\xi^2+22\xi+27}{48}; \\ & \qquad y \in [0.5-1],\ \xi \in [-1, 1] \end{aligned} \tag{3.27}$$

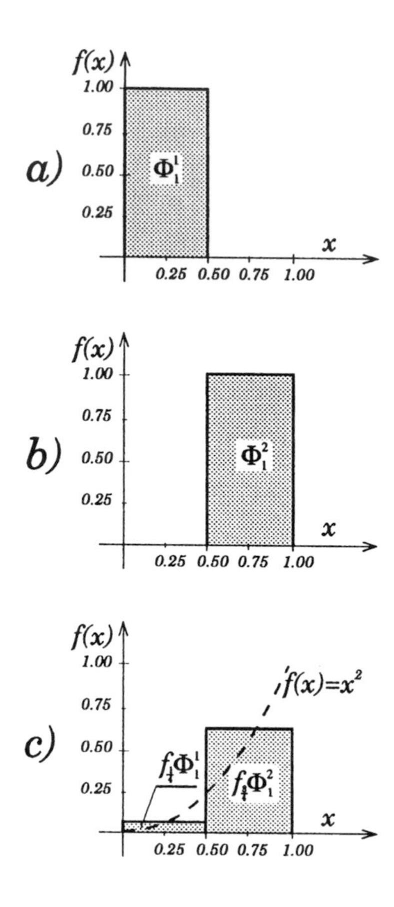

Figure 3.2: Approximation of function $f(x) = x^2$ using constant shape functions; *a*) shape function corresponding to the first interval, *b*) shape function corresponding to the second interval, *c*) approximation of the function.

Collocation requires that residuum (3.26) vanishes for $y = 0.25$, which by virtue of Eq. (3.25) corresponds to the local coordinate $\xi = 0$. Similarly residuum (3.27) should vanish for $y = 0.75$ which again corresponds to local coordinate $\xi = 0$. The final set of equations has a form

$$\begin{bmatrix} 1 & -\frac{1}{4} \\ \frac{1}{4} & 1 \end{bmatrix} \begin{bmatrix} f_{\frac{1}{4}} \\ f_{\frac{3}{4}} \end{bmatrix} = \begin{bmatrix} -\frac{5}{48} \\ \frac{9}{16} \end{bmatrix} \tag{3.28}$$

Galerkin process requires integration of the product of Eq. (3.26) and shape function (3.23) as well as the product of Eq. (3.27) and shape function (3.24). Equating the results of these integrations to zero yields a set of linear equations

$$\begin{bmatrix} \frac{1}{2} & -\frac{1}{8} \\ \frac{1}{8} & \frac{1}{2} \end{bmatrix} \begin{bmatrix} f_{\frac{1}{4}} \\ f_{\frac{3}{4}} \end{bmatrix} = \begin{bmatrix} -\frac{1}{24} \\ \frac{7}{24} \end{bmatrix} \tag{3.29}$$

The results obtained by collocation and Galerkin techniques are compared with the exact solution in Table 3.1. The same example has been solved using linear shape

Table 3.1: Solutions of integral equation obtained using two constant shape functions. Values at $x = 0.25$ and $x = 0.75$ computed by collocation (i) and Galerkin(ii).

Method	$f_{\frac{1}{4}}$	$f_{\frac{3}{4}}$
i	0.0343	0.5539
ii	0.0588	0.5686
Exact	0.0625	0.5625

functions . The integration domain has been, as previously, subdivided into two equal elements. Nodes have been placed at $x = 0$, $x = 0.5$ and $x = 1$. Corresponding unknown function values have been denoted as f_0, $f_{\frac{1}{2}}$ and f_1, respectively. Using linear shape functions (3.16) (*cf* Fig. 3.3) the variation of the unknown function can be approximated in terms of a local coordinate ξ already defined. The approximation of the unknown function has a form

$$f(x) \approx f_0\Phi_1^1(x) + f_{\frac{1}{2}}[\Phi_2^1(x) + \Phi_1^2(x)] + f_1\Phi_2^2(x) \tag{3.30}$$

(note the sum of shape functions accompanying the unknown $f_{\frac{1}{2}}$).

Substitution of Eq. (3.30) into integral equation (3.1) yields two expressions for the residuum

$$r[y(\xi)] = f_0\frac{25-21\xi}{48} + f_{\frac{1}{2}}\frac{5\xi+3}{8} + f_1\frac{3\xi-7}{48} - \frac{3\xi^2+10\xi-5}{48}; \quad y \in [0-0.5],\ \xi \in [-1,1] \tag{3.31}$$

$$r[y(\xi)] = f_0\frac{3\xi+7}{48} + f_{\frac{1}{2}}\frac{5-3\xi}{8} + f_1\frac{27\xi+23}{48} - \frac{3\xi^2+22\xi+27}{48}; \quad y \in [0.5-1],\ \xi \in [-1,1] \tag{3.32}$$

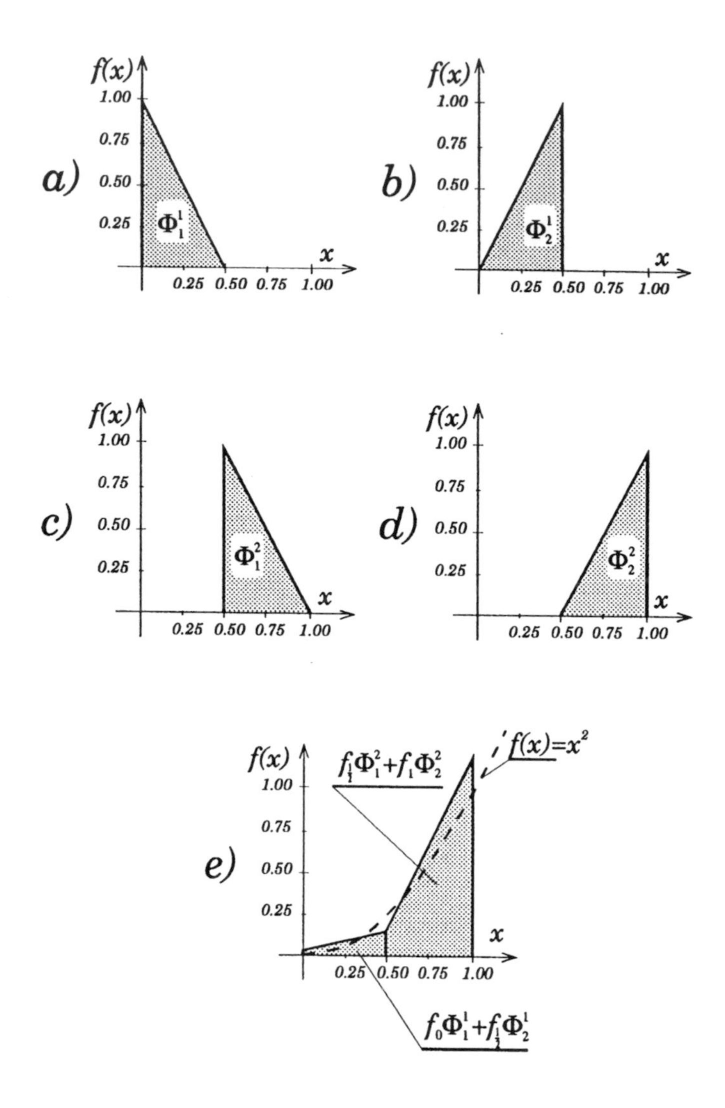

Figure 3.3: Approximation of $f(x) = x^2$ using linear shape functions; *a*), *b*) shape functions corresponding to the first interval, *c*), *d*) shape functions corresponding to the second interval, *e*) approximation of the function.

Collocation yields a condition of vanishing residua defined by Eqs. (3.31) and (3.32) at nodal points $x = 0, 0.5$ and 1. This generates a set of linear equations having a form

$$\begin{bmatrix} \frac{23}{24} & -\frac{1}{4} & -\frac{5}{24} \\ \frac{1}{12} & 1 & -\frac{1}{12} \\ \frac{5}{24} & \frac{1}{4} & \frac{25}{24} \end{bmatrix} \begin{bmatrix} f_0 \\ f_{\frac{1}{2}} \\ f_1 \end{bmatrix} = \begin{bmatrix} -\frac{1}{4} \\ \frac{1}{6} \\ \frac{13}{12} \end{bmatrix} \tag{3.33}$$

The linear equations of the Galerkin technique are produced upon integrating:

1. product of Eq. (3.31) and weighting shape function $\Phi_1^1 = (1-\xi)/2$ associated with the node $x = 0$,

2. product of Eq. (3.31) and shape function $\Phi_2^1 = (1+\xi)/2$ plus product of Eq. (3.32) and shape function $\Phi_1^2 = (1-\xi)/2$. Both weighting shape functions are associated with the node $x = 0.5$ [see Eq. (3.30)],

3. product of Eq. (3.32) and shape function $\Phi_2^2 = (1+\xi)/2$ associated with nodal point $x = 1$.

The resulting set of equations has a form

$$\begin{bmatrix} \frac{1}{6} & \frac{1}{24} & -\frac{1}{24} \\ \frac{1}{8} & \frac{1}{3} & \frac{1}{24} \\ \frac{1}{24} & \frac{1}{8} & \frac{1}{6} \end{bmatrix} \begin{bmatrix} f_0 \\ f_{\frac{1}{2}} \\ f_1 \end{bmatrix} = \begin{bmatrix} -\frac{11}{288} \\ \frac{5}{48} \\ \frac{53}{288} \end{bmatrix} \tag{3.34}$$

Solutions of Eqs. (3.33) and (3.34) are shown in Table 3.2 When comparing collo-

Table 3.2: Solutions of integral equation obtained using linear shape functions and two elements. Values at $x = 0, 0.5$ and $x = 1$ by collocation (*iii*) and Galerkin (*iv*).

Method	f_0	$f_{\frac{1}{2}}$	f_1
iii	0.0160	0.2468	0.9776
iv	-0.0417	0.2083	0.9583
Exact	0.0	0.25	1.0

cation and Galerkin approaches it should be noted that the entries of the collocation matrix are computed upon single integration whereas Galerkin entries are obtained by double integration. Thus, for the same number of degrees of freedom, forming

Galerkin matrices requires more computations. The accuracy of the Galerkin technique is usually better, but there are situations where this technique produces results less accurate than those obtained *via* collocation. Galerkin matrices can usually be made symmetric, whereas collocation does not yield symmetric matrices. On the other hand collocation based computer codes are simpler than their Galerkin counterpart. When applied to integral equations the advantages of using collocation seem to be greater than their disadvantages. This is why practically all computer BEM codes are based on this technique.

Chapter 4

Integral equation of heat conduction

4.1 Formulation of the problem

Two subsequent chapters are devoted to the application of the Boundary Element Method to heat conduction problems. The description of BEM is restricted only to those aspects of this technique that are necessary to apply BEM in heat radiation. The present chapter deals with the transformation of the original boundary value problem into an equivalent integral equation. This is, as shown in Section 1.2, the first step of the Boundary Element Method. The discretization of this integral equation comprising the next step of BEM is described in Chapter 5. The detailed discussion of the theory and the applications of BEM in various fields of engineering can be found in monographs [4, 22, 23]. The discussion of BEM within the present chapter is limited to steady state fields and simple physical situations, i.e. only the case of constant heat conductivity and homogeneous medium will be considered. Applying BEM to more complex models as *zoned media* and *nonlinear material properties* is described in the literature. The reader who wishes to study these topics is referred to the aforementioned monographs and Refs [3, 13, 15, 16, 17, 18, 19].

Steady state heat conduction taking place in a body occupying a domain V bounded by a surface S is considered. Under the above mentioned assumptions the governing equation of heat conduction has the form

$$k\nabla^2 T(\mathbf{r}) + q_V(\mathbf{r}) = 0 \quad \mathbf{r} \in V \tag{4.1}$$

where:

$T-$ temperature; K,

$k-$ heat conductivity; $W/(m \cdot K)$,

$\mathbf{r}-$ vector coordinates of the point; m,

q_V- volumetric heat generation distribution; $W/(m^3)$,

∇^2– Laplace differential operator; in 3D Cartesian coordinate system (x, y, z) this operator has an appearance:

$$\nabla^2 = \frac{\partial^2}{\partial x^2} + \frac{\partial^2}{\partial y^2} + \frac{\partial^2}{\partial z^2}$$

4.2 Reciprocity theorem

Let T^* stand for another nonzero temperature field defined within the domain V and q_V^* denote the heat sources distribution associated with this temperature field. Both functions are connected by the steady state heat conduction equation

$$k\nabla^2 T^*(\mathbf{r}) + q_V^*(\mathbf{r}) = 0 \quad \mathbf{r} \in V \tag{4.2}$$

Multiplying Eq. (4.1) by T^* and integrating the result over the entire domain V yields a result

$$\int_V T^*(\mathbf{r}) \left[k\nabla^2 T(\mathbf{r}) + q_V(\mathbf{r}) \right] dV(\mathbf{r}) = 0 \tag{4.3}$$

Obviously temperature field $T(\mathbf{r})$, being a solution of Eq. (4.1), also satisfies Eq. (4.3). The latter can also be interpreted as a weighted residuals formulation of the heat conduction equation (see Chapter 3). However, contrary to collocation and Galerkin techniques discussed in Chapter 3, the unknown function T^* is not subjected to any approximation. Thus, Eq. (4.3) is completely equivalent to the original governing equation (4.1). Integrating Eq. (4.3) twice by parts one obtains an equation known in the mathematical analysis as the *Green's identity* [48]

$$\begin{aligned} \int_V \left[T^*(\mathbf{r}) k\nabla^2 T(\mathbf{r}) - T(\mathbf{r}) k \nabla^2 T^*(\mathbf{r}) \right] dV(\mathbf{r}) &= \\ k \int_S \left[T^*(\mathbf{r}) \frac{\partial T(\mathbf{r})}{\partial n_r} - \frac{\partial T^*(\mathbf{r})}{\partial n_r} T(\mathbf{r}) \right] dS(\mathbf{r}) & \end{aligned} \tag{4.4}$$

where $\frac{\partial}{\partial n_r}$ denotes differentiation along the unit outward normal at point $\mathbf{r}$.

Let (x, y, z) denote the Cartesian coordinates of the current point $\mathbf{r}$ and (n_{rx}, n_{ry}, n_{rz}) stand for the Cartesian coordinates of the unit normal at point $\mathbf{r}$ then

$$\frac{\partial T}{\partial n_r} = \begin{cases} \frac{\partial T}{\partial x} n_{rx} + \frac{\partial T}{\partial y} n_{ry} & \text{in 2D} \\ \\ \frac{\partial T}{\partial x} n_{rx} + \frac{\partial T}{\partial y} n_{ry} + \frac{\partial T}{\partial z} n_{rz} & \text{in 3D} \end{cases} \tag{4.5}$$

Fourier's law states that the normal derivatives are proportional to *conductive heat fluxes* leaving the bounding surface, i.e.

$$q = -k\frac{\partial T}{\partial n_r} \tag{4.6}$$

$$q^* = -k\frac{\partial T^*}{\partial n_r} \tag{4.7}$$

Taking into account Eqs. (4.1), (4.2), (4.6) and (4.7) Eq. (4.4) can be rewritten in the form

$$\int_V T(\mathbf{r})q_V^*(\mathbf{r})\,dV(\mathbf{r}) = \int_S [T(\mathbf{r})q^*(\mathbf{r}) - q(\mathbf{r})T^*(\mathbf{r})]\,dS(\mathbf{r}) + \int_V q_V(\mathbf{r})T^*(\mathbf{r})\,dV(\mathbf{r}) \tag{4.8}$$

When deriving Eq. (4.8) no restrictions were imposed on the two temperature fields T, T^* arising in this equation. Hence, Eq. (4.8) is satisfied by an arbitrary pair of temperature fields. Moreover, relationship (4.8) is symmetrical with respect to these two fields. By analogy with the *Betti's reciprocity theorem* known in elasticity it is often referred to as the *reciprocity theorem.*

4.3 Fundamental solution

To transform the reciprocity theorem (4.8) into an integral equation for the unknown temperature field $T(\mathbf{r})$ the second temperature field T^* should be chosen as a known field having specific properties. The field T^* is defined as a solution of Eq. (4.2) with the internal heat source distribution taken as

$$q_V^*(\mathbf{r},\mathbf{p}) = \delta(\mathbf{p}-\mathbf{r}) \tag{4.9}$$

where δ is the Dirac's distribution [compare with Eq. (3.18)].

Function T^* satisfying, for an arbitrary pair of points: current point $\mathbf{r}$ and source point $\mathbf{p}$, Eq. (4.2) with heat source distribution defined by Eq. (4.9), is known in mathematics as the *fundamental solution* of Eq. (4.1). It should be noted that when defining the fundamental solution no boundary conditions are prescribed. The only constraint put on the temperature field T^* is an obvious requirement that the heat flux q^* should tend to zero when the distance between the *source point* $\mathbf{p}$ and the *current point* $\mathbf{r}$ tends to infinity. Fundamental solution possesses a simple physical interpretation. When viewed from this standpoint $T^*(\mathbf{r},\mathbf{p})$ is a temperature field at a point $\mathbf{r}$ being a result of heat generation from a unit point heat source acting at point $\mathbf{p}$ and placed within an infinite body. Fundamental solution can be readily obtained taking into account the symmetry of this field. The result is

$$T^*(\mathbf{r},\mathbf{p}) = \begin{cases} -\dfrac{1}{2\pi k}\ln|\mathbf{r}-\mathbf{p}| & \text{in 2D} \\[2ex] \dfrac{1}{4\pi k}\dfrac{1}{|\mathbf{r}-\mathbf{p}|} & \text{in 3D} \end{cases} \tag{4.10}$$

It should be stressed that due to the boundary conditions imposed, (heat flux prescribed) fundamental solution is not uniquely defined. The arbitrary additive constant occurring in the strict definition of the fundamental solution can be for simplicity taken as zero. The meaning of this constant has been analyzed elsewhere [50].

The heat flux q^* is also a known function and can be readily computed upon substituting Eq. (4.10) into Eq. (4.7) and keeping in mind the definition of the normal derivative (4.5). The result reads

$$q^*(\mathbf{r},\mathbf{p}) = \begin{cases} \dfrac{\mathbf{n}_r \cdot \mathbf{R}}{2\pi|\mathbf{R}|^2} & \text{in 2D} \\ \\ \dfrac{\mathbf{n}_r \cdot \mathbf{R}}{4\pi|\mathbf{R}|^3} & \text{in 3D} \end{cases} \tag{4.11}$$

where $\mathbf{n}_r \cdot \mathbf{R}$ stands for a scalar product of appropriate vectors and $\mathbf{R}$ is the vector of difference between current point $\mathbf{r}$ and source point $\mathbf{p}$ and $|\mathbf{R}|$ denotes the length of this vector.

Let (x, y, z) and (x', y', z') stand for Cartesian coordinates of the current and source point, respectively. Then the coordinates (R_x, R_y, R_z) of vector $\mathbf{R}$ has an appearance

$$\begin{aligned} R_x &= x - x' \\ R_y &= y - y' \\ R_z &= z - z' \end{aligned} \tag{4.12}$$

and the scalar product can be written as

$$\mathbf{n}_r \cdot \mathbf{R} = \begin{cases} n_{rx}(x - x') + n_{ry}(y - y') & \text{in 2D} \\ n_{rx}(x - x') + n_{ry}(y - y') + n_{rz}(z - z') & \text{in 3D} \end{cases} \tag{4.13}$$

Making use of the filtration property of Dirac's δ distribution [see Eq. (3.18)] Eq. (4.8) can be rewritten in a form

$$\begin{aligned} c(\mathbf{p})T(\mathbf{p}) &= \int_S [T^*(\mathbf{r},\mathbf{p})q(\mathbf{r}) - T(\mathbf{r})q^*(\mathbf{r},\mathbf{p})]\, dS(\mathbf{r}) \\ &- \int_V q_V(\mathbf{r})T^*(\mathbf{r},\mathbf{p})\, dV(\mathbf{r}) \end{aligned} \tag{4.14}$$

Function $c(\mathbf{p})$ assumes values between 0 and 1 depending on the placement of point $\mathbf{p}$

$$c(\mathbf{p}) = \begin{cases} 0 & \text{if } \mathbf{p} \in V \\ 1 & \text{if } \mathbf{p} \notin V \\ \in (0-1) & \text{if } \mathbf{p} \in S \end{cases} \tag{4.15}$$

For $\mathbf{p} \in S$ function c is equal to the fraction of the solid angle with vertex at $\mathbf{p}$ subtended within the body (see Fig. 4.1). Hence, for $\mathbf{p}$ located on a smooth boundary, $c = 0.5$.

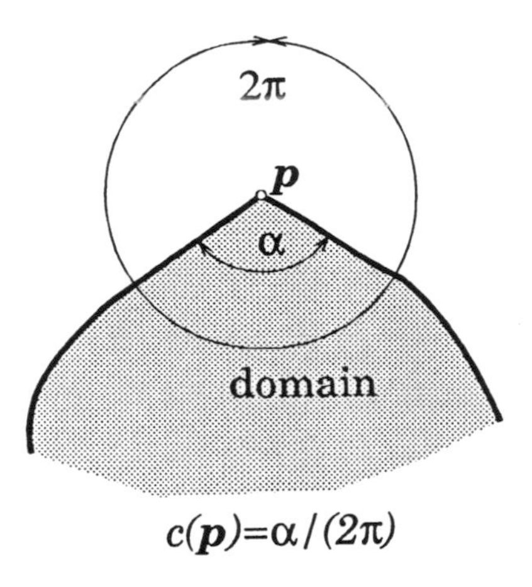

Figure 4.1: Function $c(\mathbf{p})$ at a corner point of a 2D domain.

As both T^* and q^* are known functions Eq. (4.14) could have been used to compute values of the unknown function $T(\mathbf{p})$ at arbitrary point $\mathbf{p}$ if both temperature and heat flux on the boundary were known. As long as the solution of the original equation (4.1) remains unknown, these two functions are never simultaneously known, *ie.* prescribed in boundary conditions. This will be discussed in detail in subsequent sections when analyzing the types of boundary conditions associated with the heat conduction problems.

4.4 Boundary conditions

Boundary conditions prescribe the values of temperatures and fluxes (or their interrelation) on the boundary. There are three types of linear boundary conditions encountered in heat conduction problems :

- temperature prescribed–*Dirichlet conditions,*

$$T(\mathbf{r}) = \bar{T}(\mathbf{r}) \tag{4.16}$$

- heat flux prescribed–*Neuman conditions,*

$$q(\mathbf{r}) = \bar{q}(\mathbf{r}) \tag{4.17}$$

- linear combination of temperature and flux prescribed–*Robin conditions.*

$$q(\mathbf{r}) - h(\mathbf{r})T(\mathbf{r}) = -h(\mathbf{r})T^f(\mathbf{r}) \tag{4.18}$$

The values of the functions $\bar{T}, \bar{q}, h, T^f$ entering the boundary conditions are known. Functions h and T^f occurring in Robin's condition are the heat transfer coefficients and the temperature in the core of the fluid exchanging heat with the surface of the body under consideration, respectively.

There is a great number of physical situations leading to nonlinear boundary conditions. Natural convection and radiation are typical examples of such problems. It is not possible to give a general expression encompassing all types of nonlinear conditions encountered in practice. The majority of them can be however, written in a form

$$q(\mathbf{r}) = q^n[T(\mathbf{p})] \tag{4.19}$$

where q^n is a known function of temperature.

4.5 Integral equation

Analyzing Eqs. (4.16), (4.17) and (4.18) it is clear that on a given portion of the boundary either temperature or heat flux remains unknown. Hence, Eq. (4.14) cannot be used for calculation of temperature at an arbitrary point, unless the lacking boundary values of temperature and flux are found. This can be accomplished upon placing the source point $\mathbf{p}$ on the boundary. For such a location of the source point Eq. (4.15) transforms into an integral equation (it is assumed for simplicity that the boundary surface is smooth).

$$\begin{aligned} 0.5\,T(\mathbf{p}) &= \int_S [T^*(\mathbf{r},\mathbf{p})q(\mathbf{r}) - T(\mathbf{r})q^*(\mathbf{r},\mathbf{p})]\,dS(\mathbf{r}) \\ &- \int_V q_V(\mathbf{r}')T^*(\mathbf{r}',\mathbf{p})\,dV(\mathbf{r}') \qquad \mathbf{r},\mathbf{p} \in S \quad \mathbf{r}' \in V \end{aligned} \tag{4.20}$$

In the above equation the notation of the point of integration over volume $\mathbf{r}'$ has been distinguished from the point $\mathbf{r}$ corresponding to integration over the bounding surface. Equation (4.20) is solved for:

- unknown temperature at locations where Neuman, Robin or nonlinear boundary conditions (4.19) are prescribed;
- unknown heat flux at location where Dirichlet conditions are prescribed.

Once this is done, temperature at an arbitrary point situated within the body can be computed from the equation

$$\begin{aligned} T(\mathbf{p}) &= \int_S [T^*(\mathbf{r},\mathbf{p})q(\mathbf{r}) - T(\mathbf{r})q^*(\mathbf{r},\mathbf{p})]\,dS(\mathbf{r}) \\ &- \int_V q_V(\mathbf{r}')T^*(\mathbf{r}',\mathbf{p})\,dV(\mathbf{r}') \qquad \mathbf{r} \in S \quad \mathbf{p},\mathbf{r}' \in V \end{aligned} \tag{4.21}$$

derived from Eq. (4.14) upon putting $c(\mathbf{p}) = 1$.

Equation (4.21) expresses the temperature at internal points in terms of three functions: boundary temperatures, boundary heat fluxes and internal heat generation rate. After Eq. (4.20) is solved these functions are known and Eq. (4.21) can be used to determine temperatures at an arbitrary internal point. It should be stressed that Eq. (4.21) is explicit in temperatures at internal points, thus it should be considered as an integral representation of these temperatures rather than an integral equation. Therefore, computing temperatures at internal points can be accomplished in the postprocessing phase of BEM analysis without the need for solving any equation.

Heat fluxes within the domain are seldom of interest. If necessary they can be computed from Eq. (4.21) differentiated with respect to the coordinates (x', y', z') of the source point. The x coordinate of the heat flux vector can be calculated from the equation

$$\begin{aligned} q_x(\mathbf{p}) &= -k \int_S \left[\frac{\partial T^*(\mathbf{r},\mathbf{p})}{\partial x'} q(\mathbf{r}) - \frac{\partial q^*(\mathbf{r},\mathbf{p})}{\partial x'} T(\mathbf{r}) \right] dS(\mathbf{r}) \\ &+ k \int_V q_V(\mathbf{r}') \frac{\partial T^*(\mathbf{r}',\mathbf{p})}{\partial x'} \, dV(\mathbf{r}') \qquad \mathbf{r} \in S \quad \mathbf{p}, \mathbf{r}' \in V \end{aligned} \tag{4.22}$$

where the x coordinate of the heat flux vector is defined as

$$q_x = -k \frac{\partial T}{\partial x} \tag{4.23}$$

Other components of the heat flux vector are computed analogously. As in the case of internal temperatures, internal heat fluxes can be evaluated explicitly.

In most practical situations the internal heat generation term can be neglected. In this case the integral equation of heat conduction (4.20) simplifies to

$$0.5\, T(\mathbf{p}) = \int_S [T^*(\mathbf{r},\mathbf{p}) q(\mathbf{r}) - T(\mathbf{r}) q^*(\mathbf{r},\mathbf{p})] \, dS(\mathbf{r}) \quad \mathbf{r}, \mathbf{p} \in S \tag{4.24}$$

Temperatures at internal points are computed from

$$T(\mathbf{p}) = \int_S [T^*(\mathbf{r},\mathbf{p}) q(\mathbf{r}) - T(\mathbf{r}) q^*(\mathbf{r},\mathbf{p})] \, dS(\mathbf{r}) \quad \mathbf{r} \in S \quad \mathbf{p} \in V \tag{4.25}$$

An interesting feature of Eqs. (4.24) and (4.25) is that they are formulated only on the surface bounding the domain. Hence, when compared with the original differential equation (4.1) the dimensionality of the problem has decreased by one. This has an important implication when solving numerically the integral equation (4.24); instead of discretizing the entire domain, it is enough to discretize solely the boundary of the domain. This leads to substantial time saving both in processing and during data preparation. Reduction of the dimensionality is therefore considered the main advantage of BEM.

The presence of internal heat sources causes that a volume integration should be carried out. As this spoils the elegance of BEM many efforts have been devoted to avoid volume integration. Two techniques of transforming the volume integral into a surface one have been developed. The first, referred to as *Dual Reciprocity* [65], consists in interpolation of the heat generation function by a finite number of functions $m_j(\mathbf{r})$. These functions are chosen in such a way that an equation

$$\nabla^2 \hat{T}_j(\mathbf{r}) = m_j(\mathbf{r}) \tag{4.26}$$

can be solved analytically for $\hat{T}_j$. Owing to this property of the interpolating functions m_j and their associates $\hat{T}_j$, integration by parts of the volume integral transforms it into a sum of the surface ones. In the second approach termed *Multiple Reciprocity* [63] the volume integral is approximated by a truncated series of surface integrals. Detailed discussion of these techniques is beyond the scope of this book.

Chapter 5

Discretization of heat conduction equation

5.1 Discretization of function

Integral equations derived in the preceding chapter can be solved analytically only in a very limited number of simple shaped bodies. Nontrivial cases can be handled only when a discretization procedure is employed. For the sake of simplicity the absence of internal heat generation will be first assumed.

The standard discretization technique used in BEM context is the local interpolation of the involved functions and collocation at a set of nodal points. This is a specific case of the WRM technique whose general principles have been discussed in Chapter 3. The aim of the present chapter is to describe the application of that variant of the WRM to the discretization of 2D and 3D integral equations of heat conduction. The first step of this approach is a subdivision of the integration surface S into a number of surface elements $\Delta S_e;\ e = 1, 2, \ldots, E$. Within each element both the temperature T^e and the heat flux q^e are approximated using their values at nodes $\mathbf{r}_i$ and locally based shape functions Φ_i^e.

$$T^e(\mathbf{r}) \approx \sum_{i=1}^{I^e} T_i \Phi_i^e(\mathbf{r}); \quad \mathbf{r} \in \Delta S_e \tag{5.1}$$

$$q^e(\mathbf{r}) \approx \sum_{i=1}^{I^e} q_i \Phi_i^e(\mathbf{r}); \quad \mathbf{r} \in \Delta S_e \tag{5.2}$$

where:

T_i, q_i– values at nodal point $\mathbf{r}_i$ of temperature and heat flux, respectively,

Φ_i^e– shape function associated with nodal point $\mathbf{r}_i$. This function vanishes outside the element Δ_e and at all nodal points placed within this element distinct from $\mathbf{r}_i$ where the function assumes a value of 1 [see Eq. (3.12)],

I^e– number of interpolation nodes $\mathbf{r}_i$ belonging to the element ΔS_e.

Shape functions are typically expressed in terms of normalized local coordinates. The coordinates defined within a curvilinear and surface elements are depicted in Fig 5.1. It should be noted that in the absence of internal heat generation only the boundary of the domain undergoes discretization. Hence, the dimensionality of the shape functions is one lower than that of the problem itself. Thus 1D shape functions are encountered in 2D problems whereas 2D shape functions are characteristic of 3D BEM analysis.

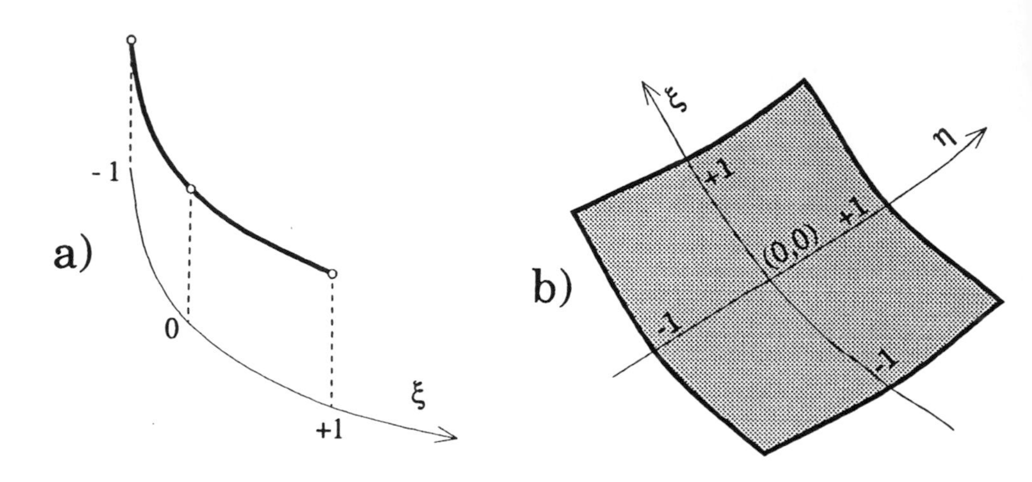

Figure 5.1: Local coordinates in boundary elements for *a)* 2D analysis, *b)* 3D analysis.

Shape functions used in 2D BEM analysis are Eqs. (3.15) to (3.17) introduced when solving 1D integral equations. The shape functions for a 3D situation can be readily constructed. Below are some examples of such functions

If one nodal point ($I^e = 1$) is placed on the element, the node is usually centered. Thus, the shape function is, in this case, associated with a point of the local coordinates $\xi = 0, \eta = 0$. The shape function is then defined as

$$\Phi_1^e(\xi,\eta) = \begin{cases} 1 & \text{if } \xi,\eta \in \Delta S_e \\ 0 & \text{otherwise} \end{cases} \tag{5.3}$$

Linear approximation of the function is accomplished upon placing four nodes ($I^e = 4$) at each corner of the element. Thus the shape functions are associated with nodes of coordinates $\xi = +1$, $\xi = -1$, $\eta = +1$ and $\eta = -1$. Functions and the local coordinates of the corresponding nodal points are listed below.

$$\begin{aligned} \Phi_1^e(\xi,\eta) &= (1-\xi)(1-\eta)/4; \text{ node } \xi = -1,\ \eta = -1 \\ \Phi_2^e(\xi,\eta) &= (1+\xi)(1-\eta)/4; \text{ node } \xi = +1,\ \eta = -1 \\ \Phi_3^e(\xi,\eta) &= (1+\xi)(1+\eta)/4; \text{ node } \xi = +1,\ \eta = +1 \\ \Phi_4^e(\xi,\eta) &= (1-\xi)(1+\eta)/4; \text{ node } \xi = -1,\ \eta = +1 \end{aligned} \tag{5.4}$$

Figure 5.2 shows a shape function belonging to the family of linear shape functions.

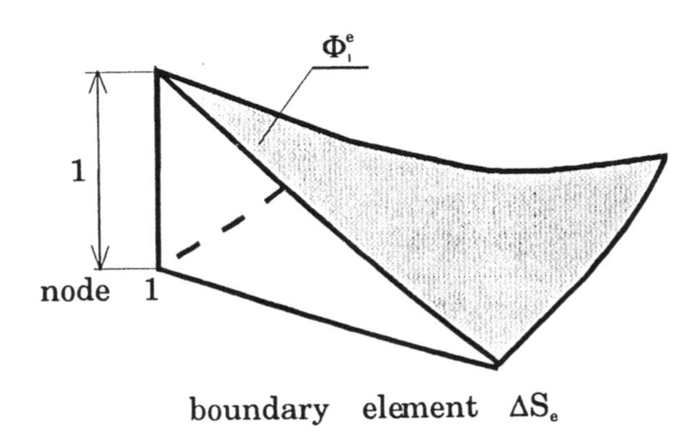

Figure 5.2: Linear shape functions defined over a curvilinear quadrangle boundary element.

Quadratic approximation of the function is accomplished upon placing eight nodes ($I^e = 8$) at the corners and midsides of the element. Appropriate formulae read

$$
\begin{aligned}
\Phi_1^e(\xi,\eta) &= (1-\xi)(1-\eta)(-\xi-\eta-1)/4; & \text{node } \xi=-1,\ \eta=-1\\
\Phi_2^e(\xi,\eta) &= (1+\xi)(1-\eta)(+\xi-\eta-1)/4 & \text{node } \xi=+1,\ \eta=-1\\
\Phi_3^e(\xi,\eta) &= (1+\xi)(1+\eta)(+\xi+\eta-1)/4 & \text{node } \xi=+1,\ \eta=+1\\
\Phi_4^e(\xi,\eta) &= (1-\xi)(1+\eta)(-\xi+\eta-1)/4 & \text{node } \xi=-1,\ \eta=+1\\
\Phi_5^e(\xi,\eta) &= (1-\xi^2)(1-\eta)/2 & \text{node } \xi=0,\ \eta=-1\\
\Phi_6^e(\xi,\eta) &= (1-\xi^2)(1+\eta)/2 & \text{node } \xi=0,\ \eta=+1\\
\Phi_7^e(\xi,\eta) &= (1-\xi)(1-\eta^2)/2 & \text{node } \xi=-1,\ \eta=0\\
\Phi_8^e(\xi,\eta) &= (1+\xi)(1-\eta^2)/2 & \text{node } \xi=+1,\ \eta=0
\end{aligned}
\tag{5.5}
$$

Let N stand for the total number of nodal points placed on the boundary. Substitution of Eqs. (5.1) and (5.2) into integral equation (4.24) and collocation at one nodal point $\mathbf{p}_k$ yields a linear equation linking nodal temperatures and heat fluxes

$$\sum_{j=1}^{N} H_{kj} T_j = \sum_{j=1}^{N} G_{kj} q_j \tag{5.6}$$

where

$$H_{kj} = \sum_e \left\{ \int_{\Delta S_e} q^*(\mathbf{r}, \mathbf{p}_k) \Phi_j^e(\mathbf{r})\, dS(\mathbf{r}) \right\} + 0.5\delta_{kj} \tag{5.7}$$

$$G_{kj} = \sum_e \left\{ \int_{\Delta S_e} T^*(\mathbf{r}, \mathbf{p}_k) \Phi_j^e(\mathbf{r})\, dS(\mathbf{r}) \right\} \tag{5.8}$$

with δ_{kj} standing for the Kronecker symbol [see Eq. (3.13)] and the summation index e running over all boundary elements sharing common node $\mathbf{r}_j$. The temperature and heat flux at this node are denoted as T_j and q_j, respectively.

When constant elements are used [shape function (5.3)] the interpolation nodes are placed at central parts of the elements. These points are obviously not shared by other elements. Thus, in this case the sum in Eqs. (5.7) and (5.8) degenerates to one term.

Equation (5.6) written for a sequence of collocation points $\mathbf{p}_k$; $k = 1, 2, \ldots, N$ constitutes a set of linear equations

$$\mathbf{HT} = \mathbf{Gq} \tag{5.9}$$

with vectors $\mathbf{T}$ and $\mathbf{q}$ containing values of temperatures and heat fluxes at collocation points. The definition of the entries of square matrices $\mathbf{H}$ and $\mathbf{G}$ is given by Eqs. (5.7) and (5.8), respectively. Matrix $\mathbf{H}$ is known as the *temperature influence matrix* whereas $\mathbf{G}$ is called the *heat flux influence matrix.*

In the presence of internal heat generation, volume integration is to be carried out. As mentioned earlier (see Section 4.5) volume integration can be converted into integration over the surface using the Dual or Multiple Reciprocity techniques. The volume integration can be also carried out in a standard manner, i.e. by subdividing the entire volume into volume cells ΔV_l; $l = 1, 2, \ldots, L$ and performing integration over these cells. Whatever method of volume integration is used, the presence of heat sources causes that set (5.9) should be modified to

$$\mathbf{HT} = \mathbf{Gq} + \mathbf{g} \tag{5.10}$$

When the standard method of handling volume integration is used the entries of the vector $\mathbf{g}$ are obtained as a sum of integrals over volume cells ΔV_l.

$$\mathbf{g}_k = \sum_{l=1}^{L} \int_{\Delta V_l} q_V(\mathbf{r}') T^*(\mathbf{r}', \mathbf{p}_k)\, dV(\mathbf{r}') \qquad \mathbf{r}' \in \Delta V_l \tag{5.11}$$

where L stands for the number of volume cells.

The integration over volume cells is usually carried out numerically upon interpolating the variation of the (known) distribution of the heat sources by shape functions. Details of this procedure can be found elsewhere [4, 23].

5.2 Approximation of the geometry

The entries of the BEM matrices are expressed as integrals over boundary elements and volume cells. These integrals can be seldom computed using analytical methods. A standard technique of evaluating these integrals is to use numerical quadratures. To accomplish this, the integral over an arbitrary shaped boundary element is transformed into an integral over a unit square. Similarly, integration over volume cells is substituted by integration over a unit cube. The (local) coordinates within each boundary element and volume cell are the same as those arising when discretizing the function (see Section 5.1). Once the transformation is carried out, the value of the appropriate integral is computed using numerical quadratures as discussed later.

The global coordinate of a current point **r** lying within a boundary element can be approximated using the concept of shape functions

$$
\begin{aligned}
x &= \sum_{i=1}^{L} x_i \Psi_i^e(\xi, \eta) \\
y &= \sum_{i=1}^{L} y_i \Psi_i^e(\xi, \eta) \\
z &= \sum_{i=1}^{L} z_i \Psi_i^e(\xi, \eta)
\end{aligned}
\tag{5.12}
$$

where x_i, y_i, z_i are the Cartesian coordinates of the nodal points defining the geometry of the element (corners, midpoint of the element sides, etc.). Symbol Ψ_i^e denotes shape functions, i.e. functions interpolating the geometry of the element between the nodal points.

Shape functions Ψ_i^e employed in Eq. (5.12) to approximate the geometry have the same properties as those used to approximate the function. Thus, functions defined by Eqs. (5.3) to (5.5) can be used to approximate the geometry of the elements. Two questions concerning the approximation of the geometry should be raised:

- constant element approximation cannot be used to represent the geometry of the surface, as using constant elements would result in discontinuity of the approximate bounding surface. This is for obvious reasons unacceptable;
- approximation orders of geometry and functions are independent. One can use, e.g. constant approximation of the function and quadratic approximation of the geometry.

The kernel q^* arising in Eqs. (4.20) to (4.25) is defined as a normal derivative of known function T^*. To perform the differentiation it is necessary to determine the unit normal vector at an arbitrary point placed within the element. The coordinates of the normal vector can be expressed in terms of derivatives of the global coordinates (x, y, z) with respect to the local coordinates ξ, η. In 3D appropriate formulae read [23]

$$
\begin{aligned}
N_{rx} &= \frac{\partial y}{\partial \xi}\frac{\partial z}{\partial \eta} - \frac{\partial z}{\partial \xi}\frac{\partial y}{\partial \eta} \\
N_{ry} &= \frac{\partial z}{\partial \xi}\frac{\partial x}{\partial \eta} - \frac{\partial x}{\partial \xi}\frac{\partial z}{\partial \eta} \\
N_{rz} &= \frac{\partial x}{\partial \xi}\frac{\partial y}{\partial \eta} - \frac{\partial y}{\partial \xi}\frac{\partial x}{\partial \eta}
\end{aligned}
\tag{5.13}
$$

where N_{rx}, N_{ry}, N_{rz} are Cartesian coordinates of the normal vector at point **r**. Similar relationships can be written for the 2D case

$$
\begin{aligned}
N_{rx} &= +\frac{\partial y}{\partial \xi} \\
N_{ry} &= -\frac{\partial x}{\partial \xi}
\end{aligned}
\tag{5.14}
$$

The unit normal vector is obtained upon dividing the coordinates (5.13) by the length of the normal vector, i.e.

$$n_{rx} = \frac{N_{rx}}{|\mathbf{N}_r|}$$
$$n_{ry} = \frac{N_{ry}}{|\mathbf{N}_r|} \tag{5.15}$$
$$n_{rz} = \frac{N_{rz}}{|\mathbf{N}_r|}$$

where $|\mathbf{N}_r|$ is the length of the normal vector defined as

$$|\mathbf{N}_r| = \sqrt{N_{rx}^2 + N_{ry}^2 + N_{rz}^2} \tag{5.16}$$

When integrating over a boundary element it is convenient to work in local coordinates (see Fig. 5.1). To accomplish this, the infinitesimal surface element $dS(\mathbf{r})$ should be expressed in the local coordinate system. It can be readily shown [23] that the surface element in local and global coordinates in 3D are connected by a relationship

$$dS(\mathbf{r}) = |\mathbf{N}_r(\xi, \eta)| d\xi d\eta \tag{5.17}$$

The analogous 2D equation reads

$$dS(\mathbf{r}) = |\mathbf{N}_r(\xi)| d\xi \tag{5.18}$$

By virtue of (5.17) integrals occurring in (5.7) and (5.8) can be written in 3D as

$$\int_{\Delta S_e} q^*(\mathbf{r}, \mathbf{p}_k) \Phi_j^e(\mathbf{r})\, dS(\mathbf{r}) =$$
$$\int_{-1}^{+1} \int_{-1}^{+1} q^*[\mathbf{r}(\xi, \eta), \mathbf{p}_k] \Phi_j^e(\xi, \eta) |\mathbf{N}_r(\xi, \eta)|\, d\xi d\eta \tag{5.19}$$
$$\int_{\Delta S_e} T^*(\mathbf{r}, \mathbf{p}_k) \Phi_j^e(\mathbf{r})\, dS(\mathbf{r}) =$$
$$\int_{-1}^{+1} \int_{-1}^{+1} T^*[\mathbf{r}(\xi, \eta), \mathbf{p}_k] \Phi_j^e(\xi, \eta) |\mathbf{N}_r(\xi, \eta)|\, d\xi d\eta \tag{5.20}$$

The 2D analogs of Eqs. (5.19) and (5.20) can be obtained straightforwardly and have a form

$$\int_{\Delta S_e} q^*(\mathbf{r}, \mathbf{p}_k) \Phi_j^e(\mathbf{r})\, dS(\mathbf{r}) = \int_{-1}^{+1} q^*[\mathbf{r}(\xi), \mathbf{p}_k] \Phi_j^e(\xi) |\mathbf{N}_r(\xi)|\, d\xi \tag{5.21}$$
$$\int_{\Delta S_e} T^*(\mathbf{r}, \mathbf{p}_k) \Phi_j^e(\mathbf{r})\, dS(\mathbf{r}) = \int_{-1}^{+1} T^*[\mathbf{r}(\xi), \mathbf{p}_k] \Phi_j^e(\xi) |\mathbf{N}_r(\xi)|\, d\xi \tag{5.22}$$

Employing a similar approach, integrals over volume cells associated with internal heat generation [see Eq. (5.11)] are transformed to integrals over unit cubes (in 3D) or unit squares (in 2D). Details of this procedure can be found in the literature [4, 23].

5.3 Integration

As already shown the entries of the BEM matrices are computed upon performing integration over boundary elements and, in the presence of internal heat generation, also over volume cells. Integration over straight 2D elements can be carried out analytically [54]. More complex elements require numerical integration.

The integrands of the integrals to be evaluated [Eqs. (5.19)– (5.22)] contain the fundamental solution T^* and its derivative q^*. These functions are singular when the distance between the integration point $\mathbf{r}$ and observation point $\mathbf{p}$ tends to zero. The singularity of the fundamental solution T^* is *weak*. It is known from elementary calculus that integrals having this type of singularity exist in common sense. Practically this means that this kind of singularity is removable. According to Eq. (4.7) function q^* is defined as the first derivative of the fundamental solution T^* thus, the singularity of q^* is more severe. This type of function behaviour is termed *strong singularity*. An integral having a strongly singular integrand do not exist in common sense, and to evaluate it a special procedure of extracting the singularity is required.

Singular behaviour of the integrals evaluated when forming BEM matrices is observed when the collocation point $\mathbf{p}_k$ belongs to the boundary element over which the integration is carried out. These integrals require special treatment. The majority of integrals arising in BEM do not exhibit singular behaviour. Such integrals are termed *regular* and can be evaluated by standard quadratures.

5.3.1 Regular integrals

Regular integrals (5.22) containing kernel function T^* are, in 2D, evaluated by employing the *Gaussian quadrature rule*

$$\int_{-1}^{+1} T^*[\mathbf{r}(\xi), \mathbf{p}_k]\Phi_j^e(\xi)|\mathbf{N}_r(\xi)|\,d\xi = \int_{-1}^{+1} F(\xi)\,d\xi \approx \sum_{n=1}^{G_\xi} w_n F(\xi_n) \tag{5.23}$$

where:

w_n– weights of the quadrature,

ξ_n– nodes of the quadrature,

G_ξ– number of quadrature nodes (abcissas) along the ξ coordinate.

Values of the weights and nodes of the Gaussian quadrature are available in the literature on BEM [4, 23] and in mathematical handbooks [1, 48, 85]. Integrals with kernel function q^* are evaluated analogously.

Integration (5.20), to be carried out in 3D BEM analysis, is accomplished upon using a 2D quadrature rule

$$\int_{-1}^{+1}\int_{-1}^{+1} T^*[\mathbf{r}(\xi,\eta),\mathbf{p}_k]\Phi_j^e(\xi,\eta)|\mathbf{N}_r(\xi,\eta)|\,d\xi d\eta =$$

$$\int_{-1}^{+1}\int_{-1}^{+1} F(\xi,\eta)\,d\xi d\eta \approx \sum_{n=1}^{G_\xi}\sum_{m=1}^{G_\eta} w_n w_m F(\xi_n,\eta_m) \tag{5.24}$$

where:

η_m– nodes of the quadrature,

G_η– number of quadrature nodes along the η coordinate.

Appropriate formulae for kernel function q^* are analogous.

The accuracy of numerical integration defined by Eqs. (5.23) and (5.24) depends on the number of nodes G_ξ, G_η used in the quadrature. It has been shown in the literature [61, 85] that the accuracy of 1D Gaussian integration is related to:

- G_ξ–number of quadrature nodes,
- s–order of singularity of the integrand,
- relative distance between the source point $\mathbf{p}_k$ and the element ΔS_e over which the integration is carried out.

Let the kernel behave like $|\mathbf{r}-\mathbf{p}|^{-s}$. Then the asymptotic formula linking the order of the quadrature and this of kernel singularity has a form [61]

$$2\frac{(2G_\xi+s-1)!}{(2G_\xi)!(s-1)}\left[\frac{D_\xi}{4L_{min}}\right]^{2G_\xi} \leq \varepsilon \tag{5.25}$$

where:

L_{min}– minimum distance between the source point $\mathbf{p}_k$ and the boundary element under consideration,

D_ξ– maximum length of the element in ξ direction,

s– order of integrand singularity,

ε– required quadrature accuracy.

The minimum distance L_{min} can be found upon solving a constrained nonlinear programming problem. An efficient method of finding this distance employing Kuhn-Tucker conditions has been addressed in Ref. [14]

Equation (5.25) enables one to determine the minimum number G_ξ of Gaussian quadrature causing that, for the given order of the singularity s, the error of integration is less than a prescribed value ε. A similar equation holds for the η axis. A procedure of adjusting the integration accuracy to the behaviour of the integrand is called *adaptive integration*. Analysis of Eq. (5.25) shows that the improvement of accuracy can be achieved either by increasing the number of quadrature nodes G_ξ (*p-adaptive integration*) or by decreasing D_ξ, the dimension of the integration domain (*h-adaptive integration*). The idea of the p -adaptation technique is obvious. This option, however, can lead to prohibitively long computing times in the case where the collocation point $\mathbf{p}_k$ lies very close to the element. This is illustrated in Fig. 5.3 taken from [54]. Inspection of this figure shows that for relative distances lower than about 0.25, increase of the number of quadrature nodes G_ξ leads to a very slow improvement of accuracy. Such cases can be readily treated using the h-adaptation where the integration domain is subdivided into subdomains. The value of the integral is obtained upon summing up contributions obtained by integrating over each of these subdomains separately.

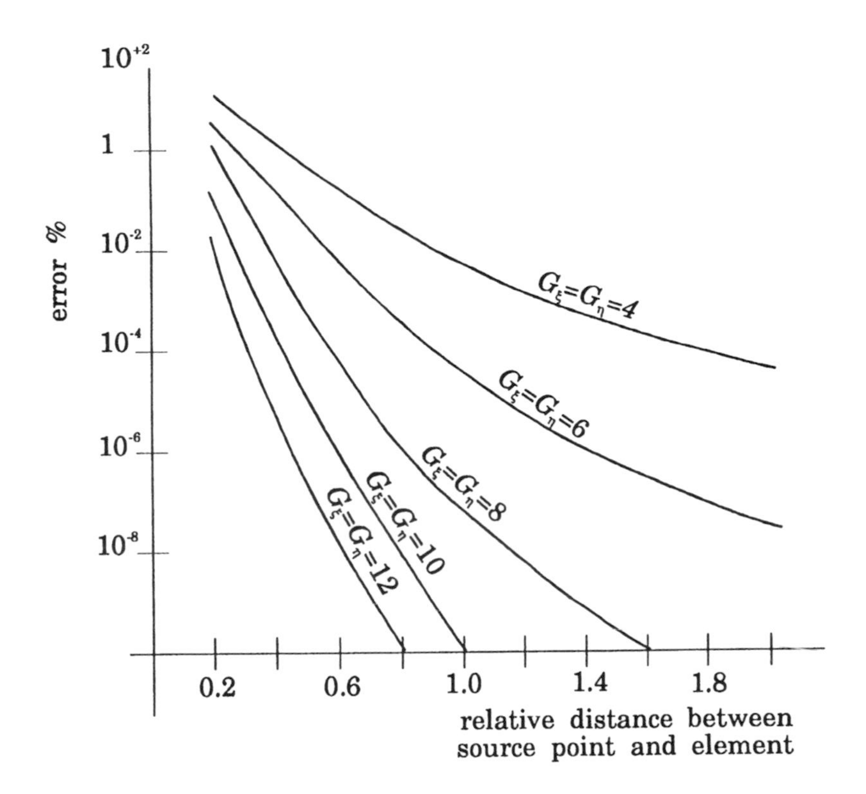

Figure 5.3: Relative error of Gaussian quadratures. Integrand singularity order $s = 2$, integration over 2D element using the same order of quadratures in both directions.

It is possible to combine these two adaptation techniques. This is often done in commercial BEM codes [14, 54] and leads to excellent accuracy within moderate computing times. Using adaptive integration is of special importance when collocation points are placed very close to a boundary element. This can happen when computing temperature fields in slender bodies as well as in the case when temperatures and heat fluxes at internal points laying in the vicinity of the bounding surface are computed.

5.3.2 Singular integrals

As mentioned, the weak singularity of the integrals can be removed. In 2D problems the weakly singular integrals can be evaluated using logarithmically weighted Gaussian quadratures [4, 23]. The difficulties associated with this type of singularity can also be circumvented by carrying out the integration in a spherical coordinate system, instead of working in Cartesian coordinates [54]. The Jacobian arising due to the change of variables causes that the singularity cancels out. Weakly singular integrals lie on the diagonal of the **G** matrix [*cf* Eqs. (5.9), (5.10)].

Strongly singular integrals contribute to the diagonal entries of the **H** matrix arising in Eq. (5.9), (5.10). Advanced techniques to handle such integrals are available in the literature [37]. They will not be discussed here, instead another simple trick known in mechanics as the *rigid body movement* will be described. Its analog in heat conduction states that within isothermal bodies all heat fluxes must vanish. This is a simple consequence of the *Second Law of Thermodynamics*. Consider Eq. (5.10). Upon putting $\mathbf{g} = \mathbf{0}$ and $\mathbf{T} = \mathbf{1}$ which is equivalent to the condition of isothermal body, one immediately obtains $\mathbf{q} = \mathbf{0}$ which in turn gives the *no flux condition*

$$\mathbf{H1} = \mathbf{0} \tag{5.26}$$

from which is follows immediately that

$$H_{kk} = - \sum_{j=1\, j \neq k}^{N} H_{kj} \tag{5.27}$$

5.4 Incorporation of boundary conditions

5.4.1 Dirichlet and Neuman conditions

In order not to introduce too many complexities at one step a situation when only Dirichlet (temperature given) and Neuman (heat flux given) boundary conditions are prescribed will be considered first.

The analysis concerns Eq. (5.10) linking temperatures and heat fluxes at nodal points belonging to the bounding surface. To solve this equation the values of temperatures and heat fluxes prescribed in boundary conditions (4.16) and (4.17) should be substituted into the vectors **T** and **q**. Putting known values to the right hand side and unknown to the left hand side, one finally arrives at a set of linear equations having a form

$$\mathbf{Kt} = \mathbf{f} \tag{5.28}$$

where the entries of the **K** matrix are defined as

$$K_{kj} = \begin{cases} H_{kj} & \text{if at point } \mathbf{r}_j \text{ heat flux is known} \\ -G_{kj} & \text{if at point } \mathbf{r}_j \text{ temperature is given} \end{cases} \tag{5.29}$$

coefficients of the vector of unknowns $\mathbf{t}$ are

$$t_k = \begin{cases} q_k & \text{if at point } \mathbf{r}_k \text{ temperature is given} \\ T_k & \text{if at point } \mathbf{r}_k \text{ heat flux is given} \end{cases} \tag{5.30}$$

and the coefficients of the right hand side vectors are computed as

$$f_k = g_k - \sum_{i_T} H_{ki_T}\bar{T}_{i_T} + \sum_{i_q} G_{ki_q}\bar{q}_{i_q} \tag{5.31}$$

where $\bar{T}_{i_T}$ and $\bar{q}_{i_q}$ denote the prescribed nodal values of temperature and heat flux, respectively. g_k stands for contribution of the internal heat source and is computed, e.g. from Eq. (5.11). The first summation in Eq. (5.30) runs over all nodes i_T with temperature prescribed. The second summation runs over nodes i_q with heat fluxes prescribed.

Solution of Eq. (5.28) for the unknown vector $\mathbf{t}$ yields the lacking temperatures and heat fluxes on the boundary.

5.4.2 Other boundary conditions

Boundary conditions other than Dirichlet and Neuman type can usually be written in a form explicitly solved for the heat flux [see Eq. 4.19]

$$q = q^n(T) \tag{5.32}$$

The Robin boundary condition [Eq. (4.18)] is a specific case of Eq. (5.32) with function $q^n(T)$ being linear in temperature. Robin's boundary condition can be used to eliminate from Eq. (5.10) the heat fluxes at nodes where this kind of condition is prescribed. Provided at other nodes Dirichlet or Neuman boundary conditions are prescribed, the matrices arising in Eq. (5.28) should be redefined as

$$K_{kj} = \begin{cases} H_{kj} & \text{if at point } \mathbf{r}_j \text{ heat flux is known} \\ -G_{kj} & \text{if at point } \mathbf{r}_j \text{ temperature is given} \\ H_{kj} - h_j G_{kj} & \text{if at point } \mathbf{r}_j \text{ Robin condition is prescribed} \end{cases} \tag{5.33}$$

with h_j denoting heat transfer coefficient at node $\mathbf{r}_j$.

Coefficients of the vector of unknowns $\mathbf{t}$ are

$$t_k = \begin{cases} q_k & \text{if at point } \mathbf{r}_k \text{ temperature is given} \\ T_k & \text{otherwise} \end{cases} \tag{5.34}$$

and the coefficients of the right hand side vectors are computed as

$$f_k = g_k - \sum_{i_T} H_{ki_T}\bar{T}_{i_T} + \sum_{i_q} G_{ki_q}\bar{q}_{i_q} - \sum_{i_R} G_{ki_R} h_{i_R} T^f_{i_R} \tag{5.35}$$

In the above equation $T^f_{i_R}$ denotes fluid temperature corresponding to nodal point $\mathbf{r}_{i_R}$. The first summation in Eq. (5.35) runs over all nodes i_T with temperature prescribed. The second summation runs over nodes i_q with heat fluxes prescribed and the third summation encompasses nodes i_R with Robin condition prescribed.

It can be seen that in the presence of linear boundary conditions of Dirichlet, Neuman and Robin type, the final set of algebraic equations (5.28) is linear.

General, nonlinear boundary conditions (5.32) lead to a set of nonlinear algebraic equations. Substitution of the nonlinear boundary condition into Eq. (5.10) yields a set of equations having a form

$$\mathbf{Kt} = \mathbf{f} + \mathbf{Dq}^n(T) \tag{5.36}$$

Matrices occurring in the above equation are defined as

$$K_{kj} = \begin{cases} H_{kj} & \text{if at point } \mathbf{r}_j \text{ heat flux is known or nonlinear con-} \\ & \text{dition (5.32) is prescribed} \\ -G_{kj} & \text{if at point } \mathbf{r}_j \text{ temperature is given} \\ H_{kj} - h_j G_{kj} & \text{if at point } \mathbf{r}_j \text{ Robin condition is prescribed} \end{cases} \tag{5.37}$$

$$t_k = \begin{cases} q_k & \text{if at point } \mathbf{r}_k \text{ temperature is given} \\ T_k & \text{otherwise} \end{cases} \tag{5.38}$$

The coefficients of the vector $\mathbf{f}$ are defined, as in the linear case, by Eq. (5.35). Entries of the matrix $\mathbf{D}$ are

$$D_{kj} = \begin{cases} G_{kj} & \text{if at point } \mathbf{r}_j \text{ nonlinear condition (5.32) is prescribed} \\ 0 & \text{otherwise} \end{cases} \tag{5.39}$$

Vector $\mathbf{q}^n(T)$ contains values of the heat flux at nodal points where nonlinear boundary condition (5.32) is prescribed. It can be readily seen that the matrix $\mathbf{D}$ is rectangular.

Due to the inherent nonlinearity of set (5.36) it can be solved only upon employing iterative solvers.

The nonlinear boundary condition (5.32) can usually be written as a sum of a linear and nonlinear functions.

$$q = h(T - T^f) + q^r(T) \tag{5.40}$$

Such form of the nonlinear boundary condition arises, e.g. when a surface exchanges heat both by convection [convective heat flux $q_c = h(T - T^f)$] and radiation [radiative heat flux $q^r(T)$].

With the nonlinear boundary condition of a form (5.40) the general appearance of the set of nonlinear algebraic equations (5.36) remains unchanged. The changes due to the new form of the nonlinear boundary condition concern only the entries of matrix $\mathbf{K}$ and vectors $\mathbf{q}^n$, $\mathbf{f}$. Matrix $\mathbf{K}$ and the right hand side vector $\mathbf{f}$ are redefined in such a way that at nodes subjected to nonlinear condition (5.40), the boundary condition of Robin type were prescribed. Vector $\mathbf{q}^n$ contains the nodal values of the nonlinear heat flux $q^r(T)$.

5.5 Internal temperatures and fluxes

Consider all nodal points located on the boundary S. Some temperatures and heat fluxes at these points are known as they are prescribed in the boundary conditions. Remaining temperatures and heat fluxes are obtained from the solution of the linear (5.28) or nonlinear (5.36) set of algebraic equations resulting from the discretization of the integral equation of heat conduction. With these nodal values known, temperature at an arbitrary point laying within the domain V can be computed upon discretizing integral equation (4.14). This can be achieved using the technique employed to discretize integral equation (4.20), i.e. using collocation and locally based shape functions. The final result of this procedure is analogous to Eq. (5.6). The temperature T_i at a chosen internal point $\mathbf{p}_i$ can be expressed as a linear combination of temperatures and heat fluxes at boundary nodal points

$$T_i = \sum_{j=1}^{N} G_{ij} q_j - \sum_{j=1}^{N} H_{ij} T_j + g_i \tag{5.41}$$

where the coefficients of this linear combination are defined analogously to Eqs. (5.7) and (5.8)

$$H_{ij} = \sum_e \left\{ \int_{\Delta S_e} q^*(\mathbf{r}, \mathbf{p}_i) \Phi_j^e(\mathbf{r}) \, dS(\mathbf{r}) \right\} \tag{5.42}$$

$$G_{ij} = \sum_e \left\{ \int_{\Delta S_e} T^*(\mathbf{r}, \mathbf{p}_i) \Phi_j^e(\mathbf{r}) \, dS(\mathbf{r}) \right\} \tag{5.43}$$

with the summation index e running over all boundary elements sharing common node $\mathbf{r}_j$. The temperature and heat flux at this node are denoted as T_j and q_j, respectively.
Term g_i gives the contribution of internal heat generation [see Eq (5.11)],

$$g_i = \sum_{l=1}^{L} \int_{\Delta V_l} q_V(\mathbf{r}') T^*(\mathbf{p}_i, \mathbf{r}') \, dV(\mathbf{r}') \qquad \mathbf{r}' \in \Delta V_l \tag{5.44}$$

When analyzing Eq. (5.41) three points should be raised:

- Eq. (5.41) is explicit in internal point temperature. Thus, to evaluate internal temperatures algebraic equations need not be solved;
- even if the boundary conditions are nonlinear, and so is the resulting set of algebraic equations, the internal temperatures are computed from a linear relationship;
- integrals over the boundary entering Eq. (5.41) are regular. However, when the point $\mathbf{p}_i$ approaches the boundary, appropriate integrals become 'almost singular' and should be evaluated using adaptive integration as discussed in section 5.3.1.

Heat fluxes at internal points can be computed upon discretizing Eq. (4.22). Appropriate relationships for the x coordinate of the heat flux vector read

$$q_x(\mathbf{p}_i) = -k \left[\sum_{j=1}^{N} G_{ij}^x q_j - \sum_{j=1}^{N} H_{ij}^x T_j + g_i^x \right] \tag{5.45}$$

where $q^x(\mathbf{p}_i)$ denote the x-coordinate of the heat flux vector at point $\mathbf{p}_i$. The coefficients occurring in the above equation are defined as

$$H_{ij}^x = \sum_e \left\{ \int_{\Delta S_e} \frac{\partial q^*(\mathbf{r}, \mathbf{p}_i)}{\partial x'} \Phi_j^e(\mathbf{r})\, dS(\mathbf{r}) \right\} \tag{5.46}$$

$$G_{ij}^x = \sum_e \left\{ \int_{\Delta S_e} \frac{\partial T^*(\mathbf{r}, \mathbf{p}_i)}{\partial x'} \Phi_j^e(\mathbf{r})\, dS(\mathbf{r}) \right\} \tag{5.47}$$

with the summation index e running over all boundary elements sharing common node $\mathbf{r}_j$.

Term g_i^x giving the contribution of internal heat generation is computed from

$$g_i^x = \sum_{l=1}^{L} \int_{\Delta V_l} q_V(\mathbf{r}') \frac{\partial T^*(\mathbf{p}_i, \mathbf{r}')}{\partial x'}\, dV(\mathbf{r}') \qquad \mathbf{r}' \in \Delta V_l \tag{5.48}$$

Remaining coordinates of the heat flux vector are computed analogously.

5.6 Symmetry of solution

Symmetry of temperature fields in heat conducting solids is frequently encountered in practical problems. One can neglect the symmetry and use this condition to check the correctness of the obtained solution. However, taking into account symmetry can lead to a substantial reduction of both computing and data preparation time. The presence of symmetry means that instead of discretizing the entire domain one can solve the problem in a smaller region. This region is to be chosen in such a way that it should be possible to construct the entire domain by summing up symmetric images of that smaller region (see Fig 5.4).

Steady state temperature distribution has a plane (3D problems) or line (2D problems) of symmetry when the domain, boundary conditions, and distribution of heat sources are symmetrical with respect to this plane (line). To simplify the terminology no distinction will be made between 2D and 3D problems and in both cases the notion of the plane of symmetry will be used.

In practical numerical computations, instead of discretizing the whole domain, it is enough to consider only its portion bounded by the symmetry planes and the original boundary (see Fig. 5.4). The boundary conditions on the original boundary remain unchanged. The symmetry of the temperature implies a zero temperature derivative on the planes of symmetry. The heat flux should therefore vanish on these planes. The symmetry planes are thus treated as if zero Neuman boundary conditions were imposed on them. This standard method of treating the symmetry reduces not only the domain to be discretized but also the number of unknowns to be found.

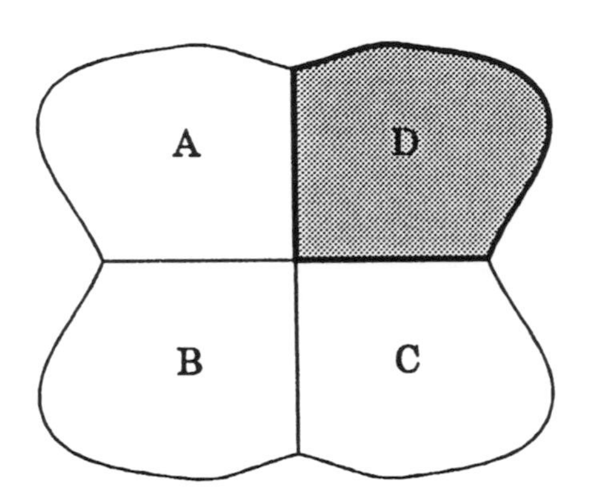

Figure 5.4: Body with two planes of symmetry. A+B+C+D– original, entire domain. A, B, C, D– symmetric portions of the domain. To solve the problem in the entire domain it is enough to find the solution in one of such portions, e.g. in subregion D.

BEM offers a more elegant method of handling symmetry [54, 57]. The technique will be described discussing the simplest possible case of a field having one plane of symmetry. The entire domain is divided by this plane into two equal parts. Superscript I and II will be appended to quantities associated with the first and second halves, respectively. Let $\mathcal{M}$ denote the symmetry relation transforming one half of the entire domain into its symmetrical counterpart. To make this abstract notion clearer consider a field with a $z = 0$ plane of symmetry. The symmetry relation is in this case defined as

$$\mathcal{M} = \begin{cases} x^I \rightarrow x^{II} \\ y^I \rightarrow y^{II} \\ z^I \rightarrow -z^{II} \end{cases} \tag{5.49}$$

The symmetry of the boundary temperatures and heat fluxes can be written as

$$T^I(\mathbf{r}) = T^{II}(\mathcal{M}\{\mathbf{r}\}) \tag{5.50}$$

$$q^I(\mathbf{r}) = q^{II}(\mathcal{M}\{\mathbf{r}\}) \tag{5.51}$$

The integral BEM equation (4.24) (the absence of internal heat generation is assumed for simplicity) can be written as

$$\begin{aligned} 0.5\,T(\mathbf{p}) &= \int_{S^I} \left[T^*(\mathbf{r},\mathbf{p})q^I(\mathbf{r}) - T^I(\mathbf{r})q^*(\mathbf{r},\mathbf{p})\right] dS^I(\mathbf{r}) \\ &+ \int_{S^{II}} \left[T^*(\mathcal{M}\{\mathbf{r}\},\mathbf{p})q^{II}(\mathcal{M}\{\mathbf{r}\}) - T^{II}(\mathcal{M}\{\mathbf{r}\})q^*(\mathcal{M}\{\mathbf{r}\},\mathbf{p})\right] dS^{II}(\mathcal{M}\{\mathbf{r}\}) \end{aligned} \tag{5.52}$$

As the fundamental solution T^* is a function of the distance between two points: $\mathbf{p}$ and $\mathbf{r}$ [see Eq.(4.10)] it is invariant with respect to the symmetry relation (see Fig 5.5)

$$T^*(\mathcal{M}\{\mathbf{r}\},\mathbf{p}) = T^*(\mathbf{r},\mathcal{M}\{\mathbf{p}\}) \tag{5.53}$$

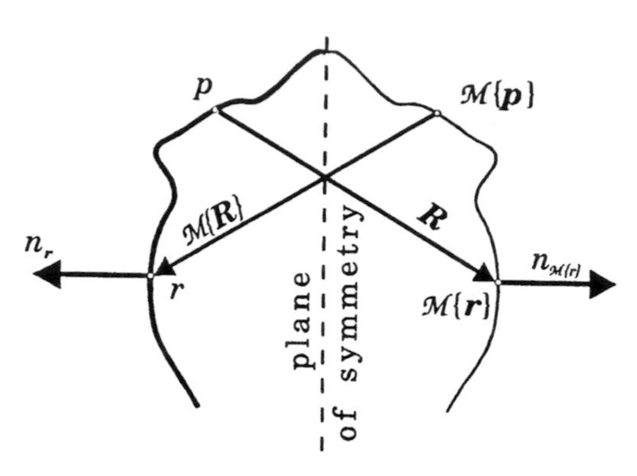

Figure 5.5: Current point $\mathbf{r}$ and source point $\mathbf{p}$ and their symmetric images $\mathcal{M}\{\mathbf{r}\}$ and $\mathcal{M}\{\mathbf{p}\}$. The distance between $\mathbf{p}$ and $\mathcal{M}\{\mathbf{r}\}$ is equal to that between $\mathbf{r}$ and $\mathcal{M}\{\mathbf{p}\}$. The angle between vectors $\mathbf{n}_{\mathcal{M}\{r\}}$ and $\mathbf{R}$ is equal to that between $\mathbf{n}_r$ and $\mathcal{M}\{\mathbf{R}\}$.

It can be shown that the heat flux q^* is also invariant with respect to the symmetry relation, i.e. it satisfies the equation

$$q^*(\mathcal{M}\{\mathbf{r}\},\mathbf{p}) = q^*(\mathbf{r},\mathcal{M}\{\mathbf{p}\}) \tag{5.54}$$

To show the correctness of Eq. (5.54) the following notation will be introduced (see Fig. 5.5):

- $\mathbf{p}$, and $\mathcal{M}\{\mathbf{p}\}$ denote the source point and its symmetric image, respectively;
- $\mathcal{M}\{\mathbf{r}\}$ and $\mathbf{r}$ denote the current point and its symmetric image, respectively (note the interchange of the symmetry relations);
- $\mathbf{n}_{\mathcal{M}\{r\}}$ and $\mathbf{n}_\mathbf{r}$ denote the normals at the current point and its symmetric image, respectively;
- $\mathbf{R}$ stands for the difference between the current point and source point, while $\mathcal{M}\{\mathbf{R}\}$ the difference between their images. Thus, vectors $\mathbf{R}$ and $\mathcal{M}\{\mathbf{R}\}$ are symmetric images of one another.

It follows from Eq. (4.11) that the heat flux q^* can be expressed in terms of the just introduced vectors as (for simplicity only 3D situation will be examined)

$$q^*(\mathcal{M}\{\mathbf{r}\},\mathbf{p}) = \frac{\mathbf{R}\cdot\mathbf{n}_{\mathcal{M}\{r\}}}{4\pi|\mathbf{R}|^3} \tag{5.55}$$

A similar relation holds with heat flux appearing on the right hand side of Eq. (5.54)

$$q^*(\mathbf{r}, \mathcal{M}\{\mathbf{p}\}) = \frac{\mathcal{M}\{\mathbf{R}\} \cdot \mathbf{n}_\mathbf{r}}{4\pi |\mathcal{M}\{\mathbf{R}\}|^3} \tag{5.56}$$

The symmetry relation does not influence the length of the vectors. Hence, the values of the denominators of Eqs. (5.55) and (5.56) are equal. Moreover, the angle between the vectors $\mathbf{n}_{\mathcal{M}\{r\}}$ and $\mathbf{R}$ is the same as that between vectors $\mathcal{M}\{\mathbf{R}\}$ and $\mathbf{n}_r$. Thus, the values of scalar products arising in the numerators are identical, which proves the correctness of Eq. (5.54).

The obvious invariance of the area of the infinitesimal surface dS with respect to the $\mathcal{M}$ transformation can be written as

$$dS^{II}(\mathcal{M}\{\mathbf{r}\}) = dS^I(\mathbf{r}) \tag{5.57}$$

Making use of this invariance, symmetry of boundary temperature (5.50), boundary heat flux (5.51) and the properties (5.53) and (5.54) of the fundamental solution Eq. (5.52) can be simplified to

$$\begin{aligned} 0.5\,T(\mathbf{p}) &= \int_{S^I} [T^*(\mathbf{r}, \mathbf{p}) + T^*(\mathbf{r}, \mathcal{M}\{\mathbf{p}\})]\, q^I(\mathbf{r}) dS(\mathbf{r}) \\ &- \int_{S^I} [q^*(\mathbf{r}, \mathbf{p}) + q^*(\mathbf{r}, \mathcal{M}\{\mathbf{p}\})]\, T^I(\mathbf{r}) dS(\mathbf{r}) \end{aligned} \tag{5.58}$$

The analysis of Eq. (5.58) leads to two conclusions:

- only half of the entire boundary should be discretized. Contrary to the standard treatment of the symmetry, the symmetry planes need not to be discretized;
- symmetry can be accounted for upon replacing the actual fundamental solution T^* by a sum of two fundamental solutions. Both solutions depend on the same current point $\mathbf{r}$. The second argument is the source point $\mathbf{p}$ and its image $\mathcal{M}\{\mathbf{p}\}$, respectively. The heat flux q^* should be modified in a similar manner. Computer implementation of the symmetry option presents no difficulties [54].

The procedure can be generalized to handle 2D problems and an arbitrary number of symmetry planes.

Chapter 6

Discretization of BEM radiation equation

6.1 Transparent medium

In order not to introduce too many complexities at one step, the basic idea of the discretization technique is first discussed for the conceptually simple transparent medium case. The radiation transfer is, in this instance, governed by Eq. (2.45). Heat conduction in the absence of internal heat sources is ruled by Eq. (4.24).

To show the similarity of the structure of these two equations, they will be recalled in a slightly rearranged form. The heat conduction equation links boundary temperature and conductive heat flux. In the absence of internal heat generation it can be rewritten as

$$0.5\,T(\mathbf{p}) = \int_S T^*(\mathbf{r},\mathbf{p})\,q(\mathbf{r})\,dS(\mathbf{r}) - \int_S q^*(\mathbf{r},\mathbf{p})\,T(\mathbf{r})\,dS(\mathbf{r}); \quad \mathbf{r},\mathbf{p} \in S \tag{6.1}$$

In the case of transparent medium the heat radiation equation links two boundary functions: the radiative heat flux and the emissive power (and hence the temperature). Analogously, as in the case of the boundary conditions associated with the heat conduction equation, to solve the heat radiation equation uniquely on portions of the boundary either emissive power or radiative heat flux should be prescribed. This problem will be addressed in Section 6.6. At this stage the interrelation of these functions will be sought for, thus they will be treated as if they were both unknown.

The equation of radiation can be written in a form similar to the conduction equation upon introducing auxiliary functions, being linear combinations of blackbody emissive power e_b and the radiative heat flux q^r. These auxiliary functions are defined as

$$\hat{T} = 2\left(\frac{q^r}{\epsilon} + e_b\right) \tag{6.2}$$

$$\hat{q} = \frac{1}{\epsilon}\left[3\epsilon e_b + q^r(3-\epsilon)\right] \tag{6.3}$$

With such definitions the heat radiation equation (2.45) can be rewritten in a form

$$0.5\,\hat{T}(\mathbf{p}) = \int_S K(\mathbf{r},\mathbf{p})\,\hat{q}(\mathbf{r})\,dS(\mathbf{r}) - \int_S K(\mathbf{r},\mathbf{p})\,\hat{T}(\mathbf{r})\,dS(\mathbf{r}); \quad \mathbf{r},\mathbf{p} \in S \tag{6.4}$$

Inspection of the definitions of the kernel functions K, T^* and q^* occurring in Eqs. (6.1) and (6.4) shows that they are singular [see Eqs. (2.35), (4.10) and (4.11)]. Moreover, as the kernel functions K and q^* are strongly singular, both equations can be classified as strongly singular integral equations. Thus, the similarity of both equations manifests itself not only in their similar appearance, but the equations of heat conduction and radiation belong to the same class of integral equations. The important consequence of this similarity is that both equations can be solved using the same numerical technique.

The standard BEM discretization method of the heat conduction equation is based on the usage of locally based shape functions and nodal collocation. Due to the already shown similarity of integral equations of heat conduction (6.1) and radiation (6.4), the idea of employing the BEM discretization technique to discretize the heat radiation equation naturally arises.

Auxiliary functions $\hat{T}$ and $\hat{q}$ have no physical interpretation and they have been introduced only for the sake of comparison of the equations of heat conduction and radiation. Hereafter, to give a better insight into the physics of the phenomenon, formulation (2.45) using the original variables: blackbody emissive power e_b and radiative heat flux q^r will be used.

With the exception of kernel function evaluation, the discretization of the heat radiation equation follows exactly the scheme described in Chapter 5, where the heat conduction equation has been dealt with. However, for the sake of completeness, the discretization of the heat radiation equation will be briefly discussed here.

The first step of the discretization procedure is the subdivision of the bounding surface S into a number of boundary elements ΔS_e. Then, the functions e_b and q^r are approximated within each element using the concept of shape functions $\Phi_i^e(\mathbf{r})$, [see Eqs. (5.1) and (5.2)].

$$e_b^e(\mathbf{r}) \approx \sum_{i=1}^{I^e} e_{bi}\Phi_i^e(\mathbf{r}); \quad \mathbf{r} \in \Delta S_e \tag{6.5}$$

$$q^{re}(\mathbf{r}) \approx \sum_{i=1}^{I^e} q_i^r\Phi_i^e(\xi,\eta); \quad \mathbf{r} \in \Delta S_e \tag{6.6}$$

where e_{bi} and q_i^r stand for nodal values of blackbody emissive power and radiative heat fluxes, respectively.

Collocation at a sequence of nodal points $\mathbf{p}_k$, with $k = 1, 2, \ldots, N$, leads to a set of linear equations linking nodal blackbody emissive powers and radiative heat fluxes. The resulting set of equations has the form

$$\mathbf{A}\mathbf{e_b}(T) = \mathbf{B}\mathbf{q}^r, \tag{6.7}$$

which is completely analogous to the discretized heat conduction equation (5.9). The entries of the matrices $\mathbf{A}$ and $\mathbf{B}$ are defined as

$$A_{kj} = -\sum_e \left\{ \int_{\Delta S_e} K(\mathbf{r}, \mathbf{p}_k) \Phi_j^e(\mathbf{r}) \, dS(\mathbf{r}) \right\} + \delta_{kj} \quad (6.8)$$

$$B_{kj} = +\sum_e \left\{ \int_{\Delta S_e} \frac{1-\epsilon(\mathbf{r})}{\epsilon(\mathbf{r})} K(\mathbf{r}, \mathbf{p}_k) \Phi_j^e(\mathbf{r}) \, dS(\mathbf{r}) \right\} - \frac{\delta_{kj}}{\epsilon(\mathbf{p}_k)} \quad (6.9)$$

In the above equations the summation index e runs over all boundary elements ΔS_e sharing common node $\mathbf{r}_j$. The blackbody emissive power and the radiative heat flux at this point will be denoted as e_{bj} and q_j^r, respectively. Vectors $\mathbf{e}_b(T)$ and $\mathbf{q}^r$ occurring in Eqs. (6.8) and (6.9) store nodal values of blackbody emissive powers and radiative heat fluxes, respectively. One should notice the similarity of the above equations with Eqs. (5.7) and (5.8) arising in heat conduction

Analysis of Eqs. (6.8) and (6.9) brings one to the conclusion that to discretize the integral equation of radiation (2.45) it is enough to evaluate the following integrals the over boundary elements

$$\int_{\Delta S_e} K(\mathbf{r}, \mathbf{p}_k) \Phi_j^e(\mathbf{r}) \, dS(\mathbf{r}) \quad (6.10)$$

The natural way of handling such integrals is to use numerical quadratures. However, this technique can only be used when the domain of integration is of simple shape, e.g. square or triangle. Usage of numerical quadratures can be extended to more complex shapes of boundary elements ΔS_e encountered in realistic problems. This can be accomplished upon projecting the original boundary element onto a unit square, resorting to the approximation of the geometry by shape functions, as discussed in Section 5.2.

Consider a 3D problem. The geometry of the boundary elements is approximated using Eqs. (5.12). The element, being in global Cartesian coordinates (x, y, z) a curvilinear quadrangle, is projected into the 2D space of local coordinates (ξ, η). The result of this projection is a unit square. With the geometry approximated by shape functions, one can readily compute from Eq. (5.15) the unit normal vector (n_{rx}, n_{ry}, n_{rz}) and from Eq. (5.17) the infinitesimal surface element. As a result, integral (6.10) is approximated in a local coordinate system by a quadrature rule analogous to Eqs. (5.19) and (5.20) of heat conduction.

$$\int_{\Delta S_e} K(\mathbf{r}, \mathbf{p}_k) \Phi_j^e(\mathbf{r}) \, dS(\mathbf{r}) \approx$$
$$\int_{-1}^{+1} \int_{-1}^{+1} K[\mathbf{r}(\xi, \eta), \mathbf{p}_k] \Phi_j^e(\xi, \eta) |\mathbf{N}_r(\xi, \eta)| \, d\xi d\eta \quad (6.11)$$

When the observation point $\mathbf{p}_k$ does not belong to the element ΔS_e over which the integration is carried out, the kernel function appearing in Eq. (6.11) is regular and appropriate integrals can be evaluated using Gaussian quadratures, as has been done in Eq. (5.24).

Singular integrals are obtained using the no flux condition (rigid body movement condition) described in Section 5.3.2. To compute the singular integrals it is assumed that the surface S is isothermal, i.e. nodal emissive powers are constant and equal, e.g. 1. In such a case the radiative heat fluxes must be zero, as heat cannot be transferred when no temperature differences are present. This leads to a condition analogous to Eq. (5.27)

$$A_{kk} = - \sum_{j=1, j \neq k}^{N} A_{kj} \tag{6.12}$$

As all off diagonal terms are expressed by regular integrals, Eq. (6.12) enables one to evaluate the diagonal term A_{kk} corresponding to the strongly singular integral. With A_{kk} known, the determination of B_{kk} presents no difficulties. In this case where the boundary elements are non concave, i.e. flat or convex, the diagonal terms need not be determined. From elementary physical interpretation, it is clear that a non concave element cannot irradiate itself. Thus, for such elements $A_{kk} = B_{kk} = 0$. This relationship can be used to test the accuracy of numerical integration, as for non concave elements the off diagonal elements must sum up to zero

$$\sum_{j=1, j \neq k}^{N} A_{kj} = 0 \text{ for non concave elements} \tag{6.13}$$

To evaluate the regular integrals the values of the cosines appearing in the definition of the kernel $K(\mathbf{r}, \mathbf{p})$ are defined as [Eq. (2.35) with $\beta = 1$ taken for simplicity]

$$K(\mathbf{r}, \mathbf{p}) = \frac{\cos \phi_r \ \cos \phi_p}{\pi |\mathbf{r} - \mathbf{p}|^2} \tag{6.14}$$

are to be determined. These cosines can be expressed as a function of the coordinates of three vectors:

$\mathbf{n}_r$ normal to the surface S at current point $\mathbf{r}$. Cartesian coordinates of this vector are denoted as (n_{rx}, n_{ry}, n_{ry});

$\mathbf{n}_p$ normal to the surface S at observation point $\mathbf{p}$. Cartesian coordinates of this vector are denoted as (n_{px}, n_{py}, n_{py});

$\mathbf{R}$ being the difference between the current and observation points $\mathbf{R} = \mathbf{r} - \mathbf{p}$. Cartesian coordinates of this vector are denoted as (R_x, R_y, R_z).

Scalar product of vector $\mathbf{R}$ and a unit normal at the observation point $\mathbf{p}$ can be written as (note the unit length of $\mathbf{n}_p$)

$$\cos \phi_p = \frac{n_{px} R_x + n_{py} R_y + n_{pz} R_z}{|\mathbf{R}|} \tag{6.15}$$

Similarly, the cosine at the current point $\mathbf{r}$ is (note the minus sign)

$$\cos \phi_r = - \frac{n_{rx} R_x + n_{ry} R_y + n_{rz} R_z}{|\mathbf{R}|} \tag{6.16}$$

Taking into account Eqs. (6.15) and (6.16) the kernel K can be expressed as

$$K(\mathbf{r},\mathbf{p}) = -\frac{(n_{rx}R_x + n_{ry}R_y + n_{rz}R_z)(n_{px}R_x + n_{py}R_y + n_{pz}R_z)}{\pi\,|\mathbf{R}|^4} \qquad (6.17)$$

The presence of shadow zones can be taken into account by adopting hidden line algorithms known in computer graphics [39]. Reference [32] contains a review of existing algorithms used in heat radiation calculations and the comparison of their numerical efficiency.

The entire discretization procedure can be extended without any difficulty to 2D problems. The differences when compared with the 3D case comprise another definition of the kernel function [Eq. (2.69)], surface element [Eq. (5.18)], and usage of 1D shape functions. The integration needed to compute the entries of matrices is to be carried out over a normalized 1D interval [-1,+1] rather than over the unit square characteristic of 3D analysis.

Example 1
Cylindrical cavity

A finite length cylindrical hole, one end of which is open to the environment, is considered [8]. The walls of the cavity are diffuse emitters and reflectors of known constant temperature and emissivity. The environment is treated as a blackbody of known temperature (radiation leaving the cavity does not return to it). This is a known benchmark example solved in Ref. [81]. The authors formulated the problem in terms of a set of two 1D integral equations solved iteratively using Simpson quadrature rules of about 100 nodal points. To check the accuracy of the BEM solution no advantage of the axial symmetry has been taken, and the problem has been solved as a 3D one. The numerical values used in computations were: temperature of the walls $T_w = 1000K$, temperature of the environment $T_e = 0K$, diameter to depth ratio $D/H = 2$. The emissivity has been varied from $\epsilon = 0.1$ to 0.9.

The ratio of the radiant energy leaving the mouth of the cavity, to the blackbody emissive power at wall temperature is called the *apparent emissivity of the cavity* and is denoted by ϵ_a. This quantity is greater than the actual emissivity of the cavity surface for arbitrary shape of the latter. This means practically, that the radiation issuing from an (isothermal) cavity always exceeds the radiant emission from a surface of the same temperature and emissivity, stretched across the opening of the cavity. In practical applications the radiative heat exchange is therefore often enhanced upon introducing artificial roughness of the surface.

Figure 6.1 shows the obtained values of apparent emissivity of cylindrical cavities for different wall emissivities. Flat rectangular constant elements have been used to model the curvilinear surface of the frustum of the cylinder. Despite these approximations the BEM results agree to within 0.001 with those obtained in Ref. [81].

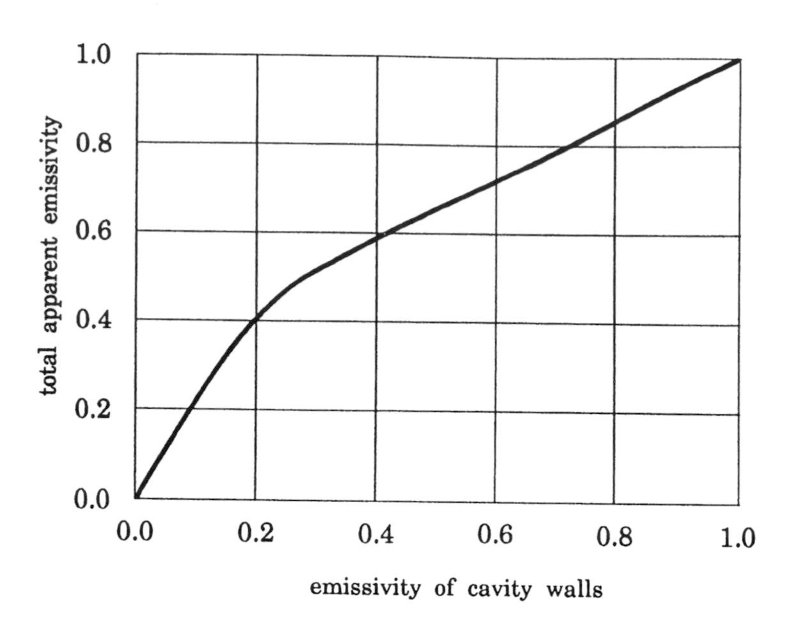

Figure 6.1: Apparent emissivity of diffusively radiating cylindrical cavity kept at a temperature of 1000 K emitting to the environment of a temperature 0 K as a function of the actual emissivity of the walls. Diameter to depth ratio equals 2. Example 1.

6.2 Isothermal, participating medium of constant absorption coefficient

Consider an enclosure filled by a homogeneous participating medium of constant temperature T^m and having an absorption coefficient a. For this specific case, the governing equations have been derived in Section 2.4.2. Their BEM formulation (2.48) and (2.49) will be recalled here for convenience.

$$
\begin{aligned}
q^r(\mathbf{p}) &+ \epsilon(\mathbf{p})e_b[T(\mathbf{p})] \\
&= \epsilon(\mathbf{p}) \int_S \left\{ e_b[T(\mathbf{r})] + \frac{1-\epsilon(\mathbf{r})}{\epsilon(\mathbf{r})} q^r(\mathbf{r}) \right\} K(\mathbf{r},\mathbf{p}) \exp\left(-a|\mathbf{r}-\mathbf{p}|\right) dS(\mathbf{r}) \\
&+ \epsilon(\mathbf{p})\, e_b(T^m) \int_S K(\mathbf{r},\mathbf{p}) \left[1 - \exp\left(-a|\mathbf{r}-\mathbf{p}|\right)\right] \, dS(\mathbf{r})
\end{aligned} \tag{6.18}
$$

$$
\begin{aligned}
q_V^r(\mathbf{p}) & + 4a\, e_b(T^m) \\
& = a \int_S \left\{ e_b[T(\mathbf{r})] + \frac{1-\epsilon(\mathbf{r})}{\epsilon(\mathbf{r})} q^r(\mathbf{r}) \right\} K_r(\mathbf{r},\mathbf{p}) \exp\left(-a|\mathbf{r}-\mathbf{p}|\right) dS(\mathbf{r}) \\
& + a\, e_b(T^m) \int_S K_r(\mathbf{r},\mathbf{p}) \left[1 - \exp\left(-a|\mathbf{r}-\mathbf{p}|\right)\right] dS(\mathbf{r})
\end{aligned}
\tag{6.19}
$$

When comparing the governing equations for transparent and participating medium the following differences should be pointed out:

- participating medium equation exists in two formulations: BEM [Eqs. (2.48), (2.49)] and standard [Eqs. (2.50) and (2.51)];
- transparent medium does not absorb radiation, thus the equation of radiative source distribution degenerates to a condition $q_V^r = 0$. To obtain the distribution of the radiative heat sources in participating medium, additional integral relationship (6.19) should be used;
- participating medium emits radiation. This is accounted for by the presence of an additional (second) integral term containing the emissive power of the medium. This term does not appear in Eq. (2.45) governing heat radiation for the transparent medium case;
- rays originating at a solid wall are attenuated along their way through the participating medium. The attenuation causes that the transmissivity function $\tau(\mathbf{r},\mathbf{p}) = \exp[-a|\mathbf{r}-\mathbf{p}|]$ arises in the first integral term. The same function occurring in the second integral term is responsible for attenuation of radiation emitted within the participating medium.

The discretization of the equations of radiation can be performed following the same path as in the case of transparent medium. The resulting set of algebraic equations corresponding to Eq. (6.18) has the form

$$
\mathbf{A}\mathbf{e}_b(T) = \mathbf{B}\mathbf{q}^r + \mathbf{c}e_b(T^m) \tag{6.20}
$$

This equation differs from those of transparent medium (6.7) by the presence of a product of a known vector $\mathbf{c}$ and the emissive power of the participating medium. This additional term is due to the radiation generated within the participating medium and is analogous to the heat source vector $\mathbf{g}$ arising in the discretized BEM equation (5.10).

Let the transmissivity between the current point $\mathbf{r}$ and the collocation point $\mathbf{p}_k$ be denoted as $\tau(\mathbf{r},\mathbf{p}_k)$ Then, the entries of $\mathbf{A}$ and $\mathbf{B}$ matrices are calculated from the relationships

$$
A_{kj} = -\sum_e \left\{ \int_{\Delta S_e} K(\mathbf{r},\mathbf{p}_k) \Phi_j^e(\mathbf{r})\, \tau(\mathbf{r},\mathbf{p}_k)\, dS(\mathbf{r}) \right\} + \delta_{kj} \tag{6.21}
$$

$$
B_{kj} = +\sum_e \left\{ \int_{\Delta S_e} \frac{1-\epsilon(\mathbf{r})}{\epsilon(\mathbf{r})} K(\mathbf{r},\mathbf{p}_k) \Phi_j^e(\mathbf{r})\, \tau(\mathbf{r},\mathbf{p}_k)\, dS(\mathbf{r}) \right\} - \frac{\delta_{kj}}{\epsilon(\mathbf{p}_k)} \tag{6.22}
$$

The entries of matrices **A** and **B** differ from those corresponding to the transparent medium case [Eqs. (6.8), (6.9)] only by the presence of transmissivity $\tau(\mathbf{r}, \mathbf{p}_k)$. The transmissivity can be, for the considered case of isothermal medium, and constant absorption coefficient, computed from Eq. (2.46) as

$$\tau(\mathbf{r}, \mathbf{p}_k) = \exp\left(-a|\mathbf{r} - \mathbf{p}_k|\right) \tag{6.23}$$

Coefficients of vector **c** are defined as

$$c_k = \sum_{j=1}^{N} \sum_{e} \int_{\Delta S_e} K(\mathbf{r}, \mathbf{p}_k) \Phi_j^e(\mathbf{r}) \left[1 - \exp\left(-a|\mathbf{r} - \mathbf{p}_k|\right)\right] dS(\mathbf{r}) \tag{6.24}$$

The equation defining the distribution of the radiative heat sources (6.19) is explicit in q_V^r function, thus it is an integral representation of the heat source rather than an integral equation. The situation here is analogous to that of computing internal temperatures and fluxes (see Section 5.5).

Discretization of Eq. (6.19) is accomplished using the already described BEM technique relying on nodal collocation and locally based shape functions. Consider a chosen internal nodal point $\mathbf{p}_i$ placed within the participating medium. The radiative source at this node, denoted as q_{Vi}^r, can be expressed as a linear combination of radiative fluxes and emissive powers at boundary nodal points. The result is analogous to Eq. (5.41) defining temperatures at internal points in the heat conduction problem

$$q_{Vi}^r = \sum_{j=1}^{N} L_{ij} q_j^r - \sum_{j=1}^{N} M_{ij} e_b(T_j) + n_i\, e_b(T^m) \tag{6.25}$$

with coefficients L_{ij}, M_{ij} defined analogously to the entries of matrices **A** and **B**, i.e. Eqs. (6.21) and (6.22).

$$L_{ij} = +a \sum_{e} \int_{\Delta S_e} \frac{1 - \epsilon(\mathbf{r})}{\epsilon(\mathbf{r})} K_r(\mathbf{r}, \mathbf{p}_i) \Phi_j^e(\mathbf{r})\, \tau(\mathbf{r}, \mathbf{p}_i)\, dS(\mathbf{r}) \tag{6.26}$$

$$M_{ij} = -a \sum_{e} \int_{\Delta S_e} K_r(\mathbf{r}, \mathbf{p}_i) \Phi_j^e(\mathbf{r})\, \tau(\mathbf{r}, \mathbf{p}_i)\, dS(\mathbf{r}) \tag{6.27}$$

where the summation index e runs over all boundary elements sharing common node $\mathbf{r}_j$ whose temperature and heat flux are denoted as T_j and q_j^r, respectively.

The term n_i in Eq. (6.25) gives the contribution of internal heat generation and corresponds to the entries of the **c** vector defined by Eq. (6.24)

$$n_i = a \sum_{j=1}^{N} \sum_{e} \int_{\Delta S_e} K_r(\mathbf{r}, \mathbf{p}_i) \Phi_j^e(\mathbf{r}) \left[1 - \exp\left(-a|\mathbf{r} - \mathbf{p}_i|\right)\right] dS(\mathbf{r}) - 4a \tag{6.28}$$

When analyzing Eq. (6.25) three points should be raised:

- Eq. (6.25) is explicit in radiative heat source function. Therefore, evaluating radiative heat sources does not require solution of algebraic equations;

- radiative heat sources can be calculated at any point placed within the participating medium. Points at which the sources are computed are independent from each other. No internal mesh needs to be introduced within the medium;

- integrals over boundary elements entering Eq. (6.25) are regular. However, when the point $\mathbf{p}_i$ approaches the boundary, appropriate integrals become 'almost singular' and should be evaluated using adaptive integration as discussed in Section 5.3.1.

Example 2
Cube of gray nonisothermal walls filled by a gray medium of constant absorption coefficient

A cube of gray diffusively radiating walls having uniform emissivity $\epsilon = 0.8$ and filled by a gray medium of absorption coefficient $a = 1\,m^{-1}$ and constant temperature 1500 K has been considered. Face z=0 of the cube was kept at temperature 300 K while the temperature of the remaining faces was maintained at 1000 K. The problem was solved using the BEM technique with 24 constant square elements.

Integration over elements was carried out using the adaptive technique described in Section 5.3.1. The calculations were performed on a personal computer. Single precision 32 bit real number representation was used in the code. With the accuracy of integration [*cf* Eq. (5.25)] set to $\varepsilon = 1 \cdot 10^{-3}$ the accuracy in satisfying the no flux condition (6.12) was of the order $1 \cdot 10^{-5}$

The computed radiative fluxes gained by elements are shown in Table 6.1. Types of elements in this Table refer to Figure 6.2, being a development of the surface of the cube. The distribution of radiative heat sources in the plane z=0.5 computed by BEM is shown in Fig 6.3.

Table 6.1: Element temperatures (given) in K and radiative fluxes (computed) in Wm^{-2} in Example 2. Type of element refers to the number shown in Fig. 6.2.

Element type	Temperature	Radiative fluxes
1	300	144.8
2	1000	93.5
3	1000	98.7
4	1000	97.6

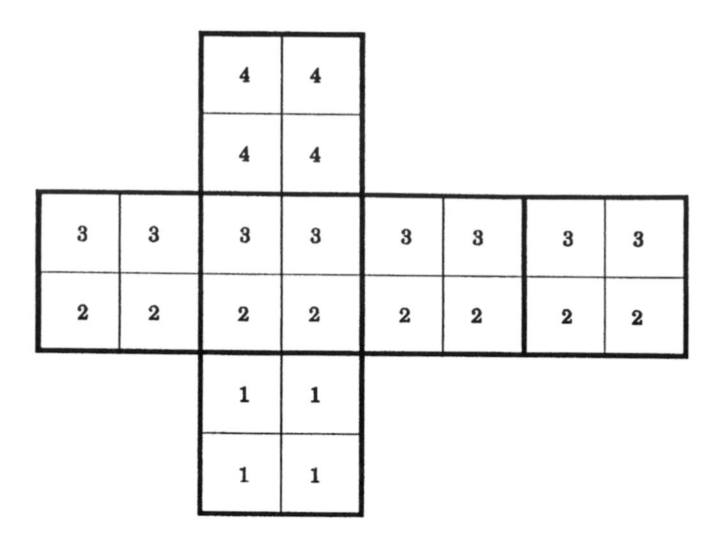

Figure 6.2: Development of the surface of the cube. Each face has been subdivided into four equal, square boundary elements. Numbers designate types of elements referred to in Table 6.1. Example 2.

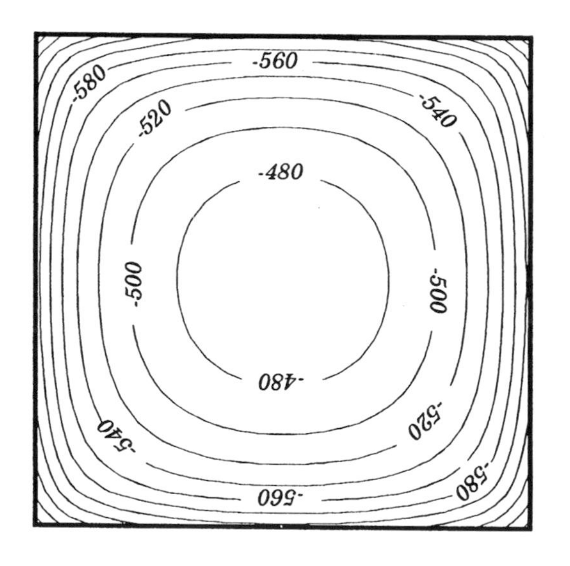

Figure 6.3: Radiative heat sources in kWm^{-3} in the plane $z = 0.5$ of the unit cube in Example 2.

6.3 Nonisothermal, gray medium

The BEM formulation of governing equations for the case of inhomogeneous spatial distribution of both temperature of the medium and absorption coefficient have been given by Eqs. (2.40) and (2.42) and will be recalled here for convenience

$$
\begin{aligned}
& q^r(\mathbf{p}) + \epsilon(\mathbf{p}) e_b[T(\mathbf{p})] \\
= \;& \epsilon(\mathbf{p}) \int_S \left\{ e_b[T(\mathbf{r})] + \frac{1-\epsilon(\mathbf{r})}{\epsilon(\mathbf{r})} q^r(\mathbf{r}) \right\} \tau(\mathbf{r},\mathbf{p}) K(\mathbf{r},\mathbf{p}) \, dS(\mathbf{r}) \\
+ \;& \epsilon(\mathbf{p}) \int_S \left\{ \int_{L_{rp}} a(\mathbf{r}') \, e_b[T^m(\mathbf{r}')] \, \tau(\mathbf{r}',\mathbf{p}) \, dL_{rp}(\mathbf{r}') \right\} K(\mathbf{r},\mathbf{p}) \, dS(\mathbf{r}) \qquad (6.29)
\end{aligned}
$$

$$
\begin{aligned}
& q_V^r(\mathbf{p}) + 4a(\mathbf{p}) e_b[T^m(\mathbf{p})] \\
= \;& a(\mathbf{p}) \int_S \left\{ e_b[T(\mathbf{r})] + \frac{1-\epsilon(\mathbf{r})}{\epsilon(\mathbf{r})} q^r(\mathbf{r}) \right\} \tau(\mathbf{r},\mathbf{p}) K_r(\mathbf{r},\mathbf{p}) \, dS(\mathbf{r}) \\
+ \;& a(\mathbf{p}) \int_S \left\{ \int_{L_{rp}} a(\mathbf{r}') \, e_b[T^m(\mathbf{r}')] \, \tau(\mathbf{r}',\mathbf{p}) \, dL_{rp}(\mathbf{r}') \right\} K_r(\mathbf{r},\mathbf{p}) \, dS(\mathbf{r}) \qquad (6.30)
\end{aligned}
$$

Equations (6.29) and (6.30) can be discretized using the already described BEM technique. The only new aspect when doing this is the presence of two line integrals

$$J_1 = \tau(\mathbf{r},\mathbf{p}) = \exp\left[-\int_{L_{rp}} a(\mathbf{r}') \, dL_{rp}(\mathbf{r}') \right] \qquad (6.31)$$

$$J_2 = \int_{L_{rp}} a(\mathbf{r}') \, e_b[T^m(\mathbf{r}')] \, \tau(\mathbf{r}',\mathbf{p}) \, dL_{rp}(\mathbf{r}') \qquad (6.32)$$

Contrary to the isothermal medium case discussed in Section 6.2 these line integrals cannot, at least in their general formulation, be evaluated analytically. However, under some additional assumptions, the analytical integration can be carried out. Many possibilities for such approximations exist, the simplest being to divide the entire volume of the participating medium into a finite number of cells and assume constant values of both temperature and absorption coefficient within each cell. Under such assumptions, integral J_1 can be calculated as (see Fig. 6.4) where for simplicity, a 2D situation is depicted]

$$J_1 = \tau(\mathbf{r},\mathbf{p}) \approx \tilde{\tau}(1, I_{rp}) = \exp\left(-\sum_{l=1}^{I_{rp}} a_l d_l \right) \qquad (6.33)$$

where:

a_l– absorption coefficient within the lth cell intersected by a ray travelling from point $\mathbf{r}$ to $\mathbf{p}$ (along the line of sight L_{rp}),

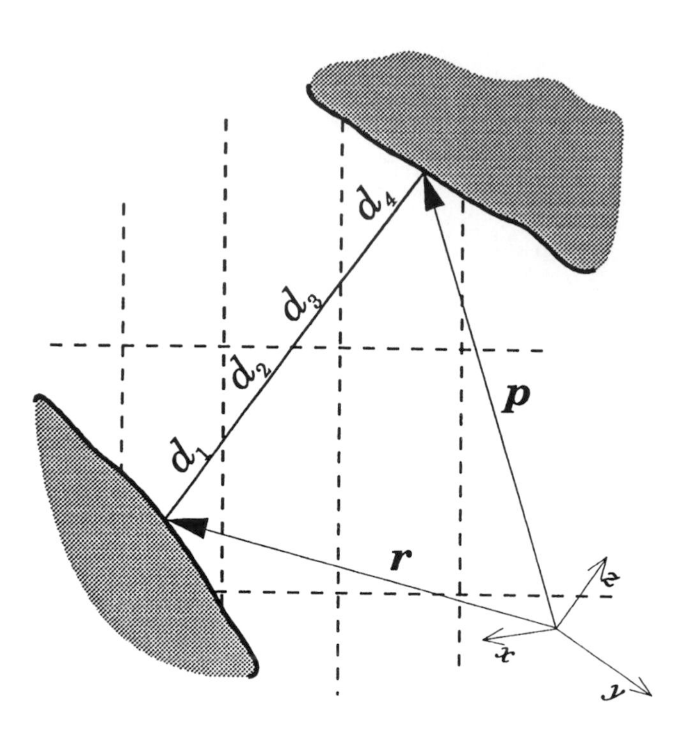

Figure 6.4: Ray intersecting volume cells. Two-dimensional situation with the number of intersected cells $I_{rp} = 4$.

d_l– length of ray within lth cell. An efficient algorithm for determining d_l is described in the Appendix,

I_{rp}– number of cells intersected by a ray travelling from point $\mathbf{r}$ to $\mathbf{p}$,

$\tilde{\tau}(1, I_{rp})$– approximate value of transmissivity between points $\mathbf{r}$ and $\mathbf{p}$.

The evaluation of the integral J_2 is somewhat more complex, but the integration can also be carried out analytically. The procedure of evaluating this integral will now be discussed in more detail.

Taking into account that the transmissivity appearing under the integral sign is also a line integral, the explicit expression for integral (6.32) can be written as

$$J_2 = \int_{L_{rp}} a(\mathbf{r}')\, e_b[T^m(\mathbf{r}')]\, \tau(\mathbf{r}', \mathbf{p})\, dL_{rp}(\mathbf{r}')$$
$$= \int_{L_{rp}} \left\{ a(\mathbf{r}')\, e_b[T^m(\mathbf{r}')] \exp\left[- \int_{L_{r'p}} a(\mathbf{r}'') dL_{r'p}(\mathbf{r}'') \right] \right\} dL_{rp}(\mathbf{r}') \quad (6.34)$$

where the symbol $\int_{L_{r'p}}(..)dL_{r'p}$ denotes integration from point $\mathbf{r}'$ to $\mathbf{p}$ along the line of sight. Points $\mathbf{r}'$ and $\mathbf{r}''$ are current points placed on the line of sight. A ray travelling from point **r** first approaches point **r**′ then point **r**″ and finally reaches point **p** (see Fig. 6.5). Let R' and R'' denote the distance between the starting point **r** of the ray

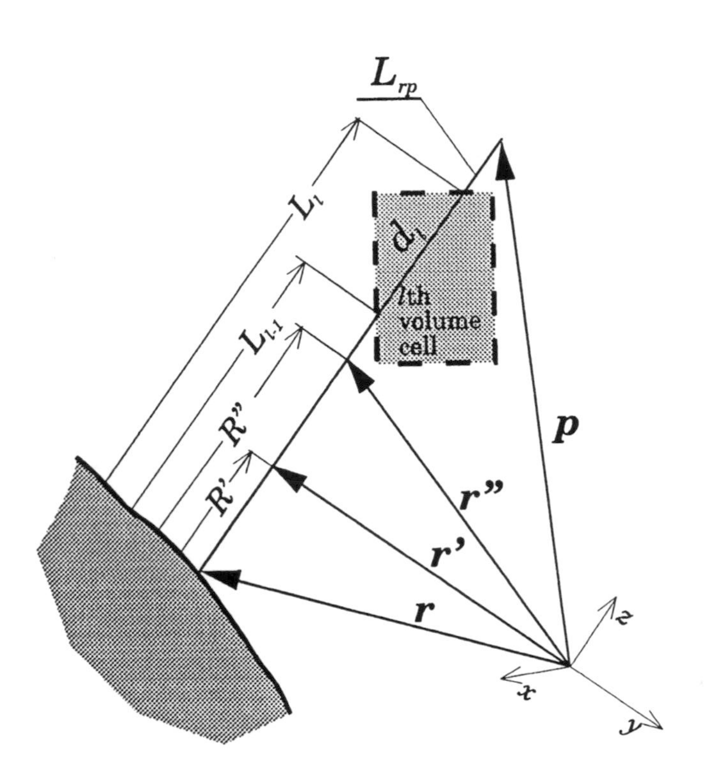

Figure 6.5: Ray travelling from point **r** to **p**.

and current points **r**′ and **r**″, respectively. The distances from point **r** and the point where the line of sight enters and leaves the lth cell will be denoted as L_{l-1} and L_l, respectively (see Fig. 6.5). With such definition L_o=0 and $L_{I_{rp}} = |\mathbf{r} - \mathbf{p}|$. The length of ray d_l satisfies the obvious relationship

$$d_l = L_l - L_{l-1}; \quad l = 1, 2, \ldots, I_{rp} \tag{6.35}$$

Taking into account that both the emissive power (temperature) and the absorption coefficient are assumed constant within a given cell, the external line integral in Eq. (6.34) can be replaced by a sum of definite integrals over d_l to give

$$J_2 \approx \sum_{l=1}^{I_{rp}} a_l e_b(T_l^m) \int_{L_{l-1}}^{L_l} \exp\left[- \int_{R'}^{L_{I_{rp}}} a(R'')\, dR'' \right] dR' \tag{6.36}$$

where T_l^m is the temperature of the participating medium in the lth cell.

The internal integral in Eq. (6.36) can be split into two parts: integral over the current cell and a sum of integrals over subsequent cells. This can be written as

$$\int_{R'}^{L_{I_{rp}}} a(R'')\,dR'' = \int_{R'}^{L_l} a_l\,dR'' + \sum_{j=l+1}^{I_{rp}} a_j d_j \tag{6.37}$$

Substitution of Eq. (6.37) into Eq. (6.36) yields

$$J_2 \approx \sum_{l=1}^{I_{rp}} a_l e_b(T_l^m)\ \exp\left(-\sum_{j=l+1}^{I_{rp}} a_j d_j\right) \int_{L_{l-1}}^{L_l} \left[\exp\left(-\int_{R'}^{L_l} a_l\,dR''\right)\right] dR' \tag{6.38}$$

After performing the integration, one arrives at

$$J_2 \approx \sum_{l=1}^{I_{rp}} e_b(T_l^m) A_l \tilde{\tau}(l+1, I_{rp}) \tag{6.39}$$

where

$$A_l = 1 - \exp(-a_l d_l) \tag{6.40}$$

This quantity has some physical interpretation. Consider a segment d_l of the line of sight belonging to the lth cell. A fraction of radiant energy emitted by points placed along this segment is absorbed by points belonging to the same segment. A_l is just this fraction and can be regarded as a measure of self absorption of radiation within a cell.

The function $\tilde{\tau}$ is defined as

$$\tilde{\tau}(l+1, I_{rp}) = \exp\left(-\sum_{j=l+1}^{I_{rp}} a_j d_j\right) \tag{6.41}$$

and denotes the transmissivity between the point where the ray leaves the lth cell and the observation point $\mathbf{p}$.

Integral J_2 can be interpreted physically as the energy that reaches the observation point $\mathbf{p}$ as a result of emission and absorption of the participating medium placed along a line of sight starting at point $\mathbf{r}$. A_l describes the attenuation within the cell of origin whereas $\tilde{\tau}(l+1, I_{rp})$ stands for the attenuation within all subsequent cells passed by the ray after it leaves the cell where energy has been emitted.

With the approximate values of line integrals determined, the discretization of the integral equation of radiation (6.29) can be readily performed by the earlier described BEM technique. The result reads

$$\mathbf{A}\mathbf{e_b}(T) = \mathbf{B}\mathbf{q}^r + \mathbf{C}\mathbf{e_b}(T^m) \tag{6.42}$$

Vector $\mathbf{e_b}(T^m)$ contains emissive powers of subsequent isothermal cells. The entries of matrices $\mathbf{A}$ and $\mathbf{B}$ are calculated from

$$A_{kj} = -\sum_e \left\{ \int_{\Delta S_e} K(\mathbf{r}, \mathbf{p}_k) \Phi_j^e(\mathbf{r}) \, \tau(\mathbf{r}, \mathbf{p}_k) \, dS(\mathbf{r}) \right\} + \delta_{kj} \tag{6.43}$$

$$B_{kj} = +\sum_e \left\{ \int_{\Delta S_e} \frac{1-\epsilon(\mathbf{r})}{\epsilon(\mathbf{r})} K(\mathbf{r}, \mathbf{p}_k) \Phi_j^e(\mathbf{r}) \, \tau(\mathbf{r}, \mathbf{p}_k) \, dS(\mathbf{r}) \right\} - \frac{\delta_{kj}}{\epsilon(\mathbf{p}_k)} \tag{6.44}$$

The similarity of Eqs. (6.43) and (6.44) and relationships defining the entries of the matrices in the isothermal case (6.21) (6.22) can be readily noticed. The only difference between these two groups of equations is another method of calculating the transmissivity. In isothermal media transmissivity is calculated from Eq. (6.23) whereas in the considered nonisothermal case the transmissivity is computed from relationship (6.33).

Coefficients of the matrix **C** are computed as

$$C_{kl} = \sum_{j=1}^{N} \sum_e \int_{\Delta S_e} K(\mathbf{r}, \mathbf{p}_k) \Phi_j^e(\mathbf{r}) \, A_l \, \tilde{\tau}(l+1, I_{rp_k}) \, dS(\mathbf{r}) \tag{6.45}$$

where L_V denotes the number of volume cells and I_{rp_k} is the number of cells intercepted by a ray originating at point $\mathbf{r}$ and reaching the collocation point $\mathbf{p}_k$

External summation in Eq. (6.45) only apparently runs over all boundary nodes. In fact from all rays arriving at the observation point $\mathbf{p}_k$ only those crossing a cell having a temperature T_l^m contribute to the entry C_{kl}. For other lines of sight the length of ray d_l intercepted by the boundaries of the cell is zero, hence A_l vanishes and the entire summation vanishes too.

Discretization of Eq. (6.30) yields

$$q_{Vi}^r = \sum_{j=1}^{N} L_{ij} q_j^r - \sum_{j=1}^{N} M_{ij} e_b(T_j) + \sum_{l=1}^{L_V} N_{il} e_b(T_l^m) \tag{6.46}$$

with coefficients L_{ij}, M_{ij} defined exactly as in the isothermal medium case, i.e. by Eqs. (6.26) and (6.27). As in Eqs. (6.43) and (6.44) the difference between the isothermal and nonisothermal case resolves itself to another formula used to calculate the transmissivity τ.

Coefficients N_{il} are computed from a relationship

$$N_{il} = a_i \sum_{j=1}^{N} \sum_e \int_{\Delta S_e} K_r(\mathbf{r}, \mathbf{p}_i) \Phi_j^e(\mathbf{r}) A_l \, \tilde{\tau}(l+1, I_{rp_i}) \, dS(\mathbf{r}) - 4\delta_{li} a_i \tag{6.47}$$

where I_{rp_i} denotes the number of cells intercepted by a ray originating at point $\mathbf{r}$ and reaching the collocation point $\mathbf{p}_i$.

6.4 Nongray, nonisothermal medium; band approximation

This model of participating medium has been discussed in Section 2.4.3. The most characteristic assumption of this model is that spectral dependence of the absorption coefficient a_λ can be approximated by a step function. The absorption coefficient vanishes within certain spectral intervals (windows) and assumes constant, nonzero values within other spectral intervals (bands).

Let u denote the number of a subsequent spectral interval (window or band). The radiative heat flux transmitted within the uth interval is denoted by q^{ru}, radiative source corresponding to this interval is denoted by q_V^{ru}. The fraction of emissive power corresponding to uth spectral interval is denoted by e_b^u. These functions defined by Eqs. (2.55)–(2.57) as integrals of their spectral analogs over spectral intervals will be recalled here for convenience

$$q^{ru}(\mathbf{r}) = \int_{\lambda^u}^{\lambda^{u+1}} q_\lambda^r(\lambda', \mathbf{r})\, d\lambda' \tag{6.48}$$

$$q_V^{ru}(\mathbf{r}) = \int_{\lambda^u}^{\lambda^{u+1}} q_{\lambda V}^r(\lambda', \mathbf{r})\, d\lambda' \tag{6.49}$$

$$e_b^u(\mathbf{r}) = \int_{\lambda^u}^{\lambda^{u+1}} e_{b\lambda}(\lambda', \mathbf{r})\, d\lambda' \tag{6.50}$$

To determine the radiative heat fluxes, sources and emissive powers transferred within subsequent spectral intervals, the equations of radiation are formulated and discretized for each spectral interval. As described in Section 2.4.3 once these quantities are known, the amount of heat transferred within the entire spectrum is computed as a sum of band and window contributions.

Heat transfer within windows and bands will be analyzed separately. Symbols a^u and ϵ^u and τ^u will stand for the values of the absorption coefficient, wall emissivity and transmissivity associated with the uth spectral interval.

6.4.1 Radiative transfer within bands

This phenomena is governed by Eqs. (2.60) and (2.61). Discretization of these equations is accomplished using the same nodal collocation technique employed when discretizing the gray model equations (see Section 6.3). The result reads

$$\mathbf{A}^u \mathbf{e_b}^u(T) = \mathbf{B}^u \mathbf{q}^{ru} + \mathbf{C}^u \mathbf{e_b}^u(T^m) \tag{6.51}$$

$$q_{Vi}^{ru} = \sum_{j=1}^{N} L_{ij}^u q_j^r - \sum_{j=1}^{N} M_{ij}^u e_b^u(T_j) + \sum_{l=1}^{L_V} N_{il}^u e_b^u(T_l^m) \tag{6.52}$$

Quantities entering Eqs. (6.51) and (6.52) are defined analogously to their counterparts arising in the gray model equations (6.42) and (6.46). Appropriate relationships are obtained upon appending superscript u to functions ϵ, a, q^r, q_V^r, e_b, τ of the gray medium model. The procedure is straightforward and does not require further explanation. For illustration purposes, only three examples of band quantities will be given

$$\{\mathbf{e_b}^u(T)\}_j = \int_{\lambda^u}^{\lambda^{u+1}} e_{b\lambda}(\lambda', \mathbf{r}_j)\, d\lambda' \tag{6.53}$$

$$B_{kj}^u = +\sum_e \left\{ \int_{\Delta S_e} \frac{1-\epsilon^u(\mathbf{r})}{\epsilon^u(\mathbf{r})} K(\mathbf{r}, \mathbf{p}_k) \Phi_j^e(\mathbf{r})\, \tau^u(\mathbf{r}, \mathbf{p}_k)\, dS(\mathbf{r}) \right\} - \frac{\delta_{kj}}{\epsilon^u(\mathbf{p}_k)} \tag{6.54}$$

$$N_{il}^u = a_i^u \sum_{j=1}^{N} \sum_e \int_{\Delta S_e} K_r(\mathbf{r}, \mathbf{p}_i) \Phi_j^e(\mathbf{r}) A_l^u\, \tilde{\tau}^u(l+1, I_{rp_i})\, dS(\mathbf{r}) - 4\delta_{li} a_i^u \tag{6.55}$$

where

$$A_l^u = 1 - \exp(-a_l^u d_l) \tag{6.56}$$

$$\tilde{\tau}^u(l+1, I_{rp}) = \exp\left(- \sum_{j=l+1}^{I_{rp}} a_j^u \right) \tag{6.57}$$

and a_l^u stands for absorption coefficient of the lth cell within uth band. Definitions of the remaining quantities entering Eqs. (6.51) and (6.52) are obvious.

6.4.2 Radiative transfer within windows

Equation (2.62) describing the energy transfer within windows is discretized by the technique used to deal with transparent medium (see Section 6.1). The discretized form of Eq. (2.62) has an appearance

$$\mathbf{A}^u \mathbf{e_b}^u(T) = \mathbf{B}^u \mathbf{q}^{ru} \tag{6.58}$$

Entries of matrices arising in Eq. (6.58) are defined in an analogous manner to their transparent medium counterparts upon appending superscript u to appropriate quantities, e.g.

$$\{\mathbf{q}^{ru}\}_j = \int_{\lambda^u}^{\lambda^{u+1}} q^{ru}(\lambda', \mathbf{r}_j)\, d\lambda' \tag{6.59}$$

$$B_{kj}^u = +\sum_e \left\{ \int_{\Delta S_e} \frac{1-\epsilon^u(\mathbf{r})}{\epsilon^u(\mathbf{r})} K(\mathbf{r}, \mathbf{p}_k) \Phi_j^e(\mathbf{r})\, dS(\mathbf{r}) \right\} - \frac{\delta_{kj}}{\epsilon^u(\mathbf{p}_k)} \tag{6.60}$$

Obviously, the same result can be achieved upon putting in band equations of Section 6.4.1 the absorption coefficient a^u equal to zero. One should also observe that as the medium is transparent, radiative heat sources vanish and need not be computed.

Example 3
Nongray nonisothermal medium filling enclosure of nongray, nonisothermal walls

The problem considered [10] is an open rectangular prism of dimensions $0.7 \times 0.7 \times 0.4$ m. The participating medium filling the cavity has nonuniform, known temperature. The temperature of the walls is known. The base of the cavity is maintained at 700 K, whereas the mouth of the cavity is open to an environment of temperature 300 K. The temperatures of other walls vary linearly in height from 700 K at the bottom to 300 K at the top of the cavity. The entire spectrum has been subdivided into five spectral intervals. Within each interval both emissivities of the wall and the absorption coefficients of the gas were assumed independent of the wavelength.

The walls of the enclosure have been subdivided into 210 square elements (0.1×0.1 m). The base of the cavity has been subdivided into three zones as depicted in Fig. 6.6. Each zone has different emissivity ϵ_1, ϵ_2, ϵ_3. These emissivities, along with the absorption coefficients and spectral interval limits, are shown in Table 6.2. Other quantities arising in this table have the following meaning: ϵ_4 stands for the emissivity of the side walls and ϵ_5 is the emissivity of the environment.

Table 6.2: Spectral intervals and associated values of material properties used in Example 3.

Interval	Wavelength μm	Emissivities					Absorption coefficient $a\, m^{-1}$
		ϵ_1	ϵ_2	ϵ_3	ϵ_4	ϵ_5	
1	0.00 – 2.60	0.9	0.1	0.1	0.5	1.0	0.0
2	2.60 – 2.86	0.7	0.3	0.2	0.5	1.0	1.0
3	2.86 – 12.6	0.5	0.5	0.3	0.5	1.0	0.0
4	12.6 – 18.5	0.3	0.7	0.2	0.5	1.0	0.1
5	18.5 – ∞	0.1	0.9	0.1	0.5	1.0	0.0

The volume of the cavity has been divided into 27 rectangular prism cells gathered in three layers of nine cells each. The heights of the bottom and middle layers were 0.1 m and that of the top layer was 0.2 m. The temperatures of the cells in subsequent layers along with their geometry are depicted in Figs. 6.7a–6.7c.

Figures 6.8–6.12 show the computed values of the radiative heat fluxes at the base of the cavity.

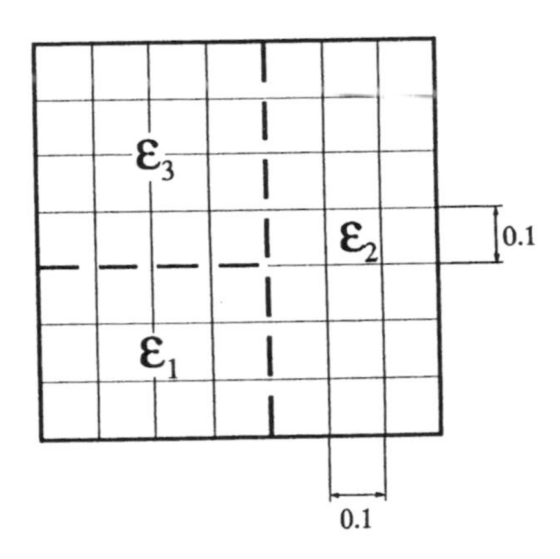

Figure 6.6: Numerical grid at the base of the cavity and the subdivision of the base into three zones of different emissivity. Example 3.

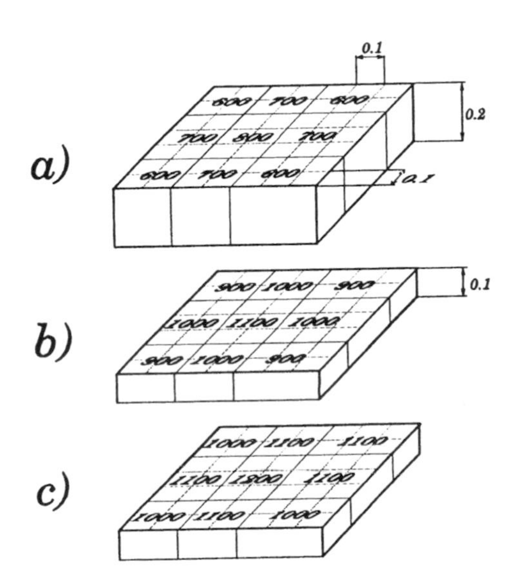

Figure 6.7: Geometry and temperatures (Kelvins) in participating medium cells in the *a)* top, *b)* middle, and *c)* upper layer of the medium. Example 3.

When analyzing the results the following points need to be mentioned:

- the shape of the emissive power *versus* wavelengths curve is such that the bulk of energy is transmitted at wavelengths in the vicinity of the maximum blackbody emissive power (see Fig. 2.4). The third interval contains the maximum of the blackbody emissive power for the temperature of the bottom. Therefore, despite the narrowness of this interval, most of the energy is transmitted within it (compare Figs. 6.8*b* and 6.11*b*);

- maximum values of the radiative fluxes for each band are concentrated in sub-regions having greater emissivity (Figs. 6.9–6.12);
- the temperature of the medium is higher than that of the base of the cavity. Thus, in intervals where the medium is participating in the radiative heat exchange, heat is transmitted from the participating medium to the base. Within the second and fourth spectral intervals the heat lost by the bottom is less than it would be if the medium were transparent (compare Figs. 6.9*a* with 6.9*b* and 6.10*a* with 6.10*b*);
- a large value of the absorption coefficient within the second band causes that the heat gained by the base of the cavity from the radiating medium is greater than the heat lost due to the radiative exchange with side walls and the environment (Fig. 6.9*b*). Within the fourth interval this effect is less pronounced because in this band the absorption coefficient is ten times smaller than in the second interval (see Figs. 6.9*b*, 6.10*b*).

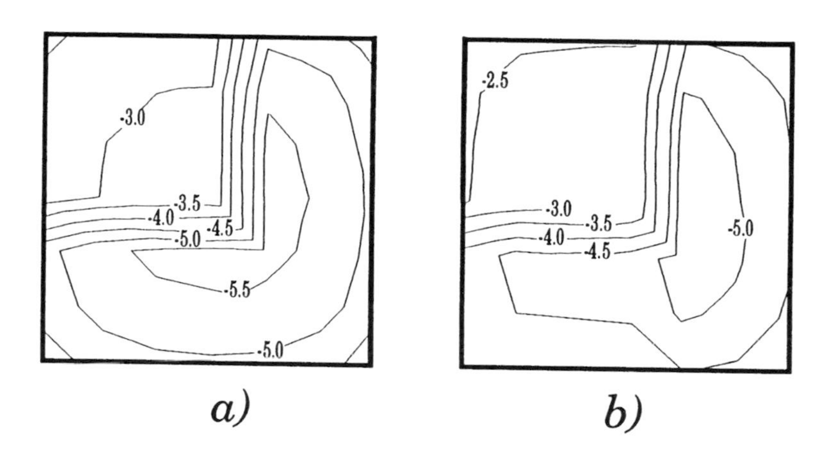

Figure 6.8: Total incoming radiative heat flux (in kWm^{-2}) at the base of the rectangular prism cavity for: *a)* transparent medium and *b)* participating medium filling the cavity. Example 3.

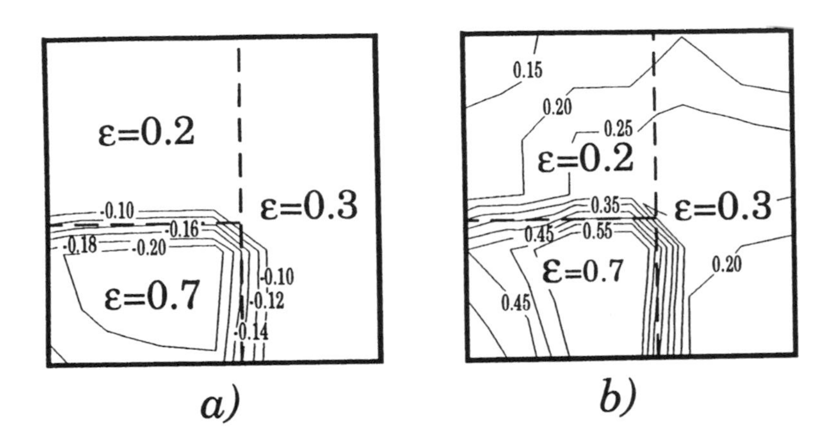

Figure 6.9: Incoming radiative heat flux (in kWm^{-2}) at the base of cavity within the second spectral interval for: *a)* transparent medium $a = 0\,m^{-1}$ and *b)* participating medium $a = 1m^{-1}$ filling the cavity. Example 3.

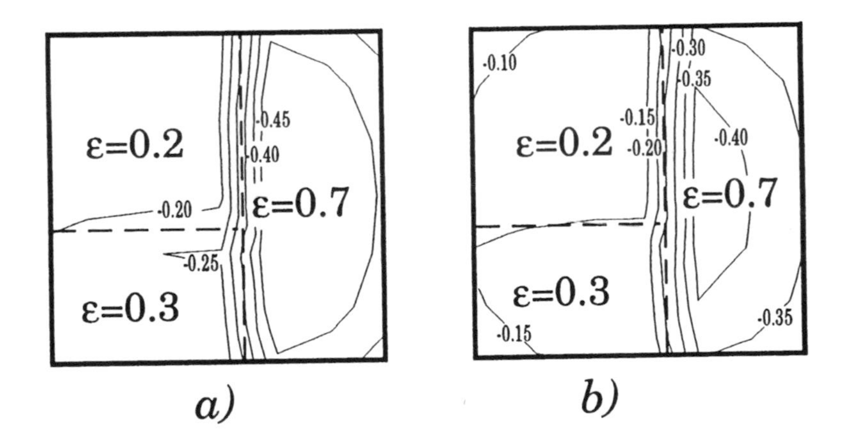

Figure 6.10: Incoming radiative heat flux (in kWm^{-2}) at the base of cavity within the fourth spectral interval for: *a)* transparent medium $a = 0\,m^{-1}$ and *b)* participating medium $a = 0.1\,m^{-1}$ filling the cavity. Example 3.

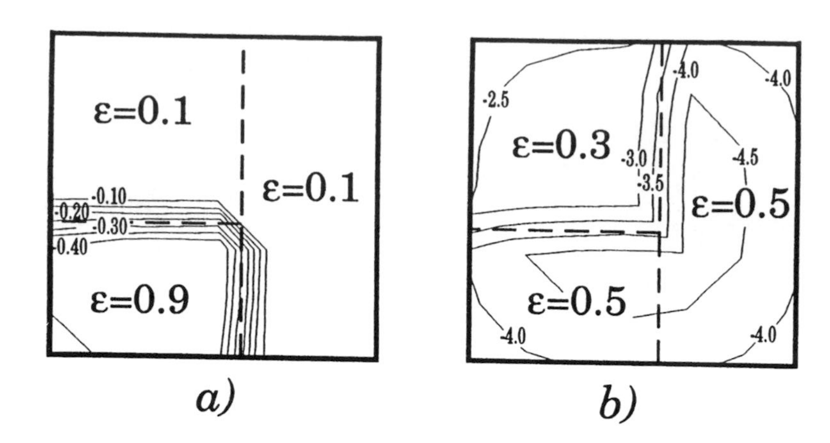

Figure 6.11: Incoming radiative heat flux (in kWm^{-2}) at the base of cavity within the *a)* first and *b)* third spectral intervals where the medium is transparent to radiation. Example 3.

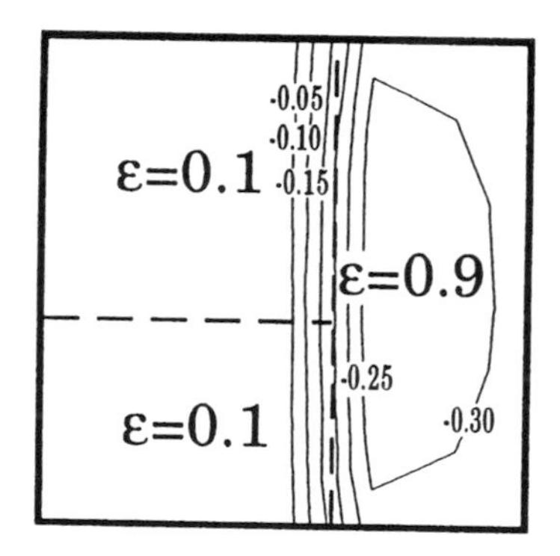

Figure 6.12: Incoming radiative heat flux (in kWm^{-2}) at the base of cavity within the fifth spectral interval where the medium is transparent to radiation. Example 3.

6.5 No flux condition for radiation in participating media

The no flux condition(rigid body movement) states that in the absence of temperature gradients heat fluxes vanish. This condition, discussed in Section 5.3.2 and employed in the transparent medium case [see Eq. (6.12)] can also be generalized for the case of radiation in participating medium both in a gray and nongray case. For the sake of simplicity only the gray nonisothermal medium model will be discussed. The results obtained here can be readily generalized for any other model of participating medium. No flux condition is useful when calculating the strongly singular integrals, arising when the collocation point lies on the boundary element over which the integration is carried out. It is clear that in the case of isothermal walls enclosing isothermal participating medium kept at wall temperature, both radiative heat fluxes and sources should vanish. This leads to two relationships linking the coefficients of BEM radiation matrices

$$\sum_{j=1}^{N} A_{kj} = \sum_{l=1}^{L_V} C_{kl} \tag{6.61}$$

$$\sum_{j=1}^{N} M_{ij} = \sum_{l=1}^{L_V} N_{il} \tag{6.62}$$

Equations (6.61) and (6.62) can be used to compute singular integrals associated with self irradiation of concave elements. When flat boundary elements are used these equations can serve as a test of the accuracy of quadratures used to compute the entries of BEM matrices.

6.6 Conditions of uniqueness

Two integral equations of heat radiation (2.40) and (2.42) connect four functions:

i. temperature (blackbody emissive power) of the bounding surface,

ii. radiative heat flux on the bounding surface,

iii. temperature (blackbody emissive power) of the participating medium,

iv. radiative heat sources within the participating medium.

To solve the problem uniquely two of these functions must be known. Not all combinations of the the prescribed functions lead to unique solutions. In this section some hints concerning the proper choice of the known functions will be given.

Radiative problems of practical importance are solved in conjunction with other heat transfer modes. The coupling conditions are energy balances on the enclosing surfaces as well as on the volume of participating medium. This problem will be dealt with in the subsequent chapter, at this stage it is enough to mention that, due to the nonlinearity introduced by the dependence of the emissive power on temperature, the numerical process of adjusting these balances is to be performed iteratively. Which of the four functions are known when the equations of radiation are solved results from the iterative process used. Conversely, the modeller can choose such an iterative technique that yields a convenient set of known functions.

The functions *i* and *ii* are defined on the boundary. At each location of the boundary either temperature or radiative heat flux should be prescribed. The functions *iii* and *iv* are defined within the entire volume of the participating medium. Again, one of these functions should be known to obtain a unique solution. Practically the most convenient combination of prescribed functions is when both the temperature of the surface and the medium are known. This set guarantees the uniqueness of the solution. The process of forming the BEM matrices can be, in this case, programmed in a straightforward manner even when the material properties are temperature dependent. Another advantage of this set of prescribed functions is that the equations are explicit in heat radiative sources, hence solution of one set of algebraic equations can be avoided.

Theoretically it is also possible to work with both functions: temperature and radiative heat flux known on the boundary or with the remaining two functions prescribed. These alternatives can lead to sets with non square matrices. Little is known about the behaviour of the solution in these cases, thus such combinations of known functions should be avoided in practical computations.

Three frequently encountered physical situations connected with the conditions of uniqueness need some explanation:

Contact of an open cavity with the environment of known temperature. As mentioned in Example 1 the radiant energy leaving the cavity practically does not return to it. In this case the mouth of the enclosure is treated as a blackbody of known temperature.

Presence of symmetry planes in the solution fields. The standard method of handling these planes in heat conduction problems, is to treat them as new boundaries with zero conductive heat flux prescribed (see Section 5.6). This approach cannot be used when solving heat radiation problems. Although the radiative heat flux vanishes on the planes of symmetry, the integral equations of radiation have been derived for diffuse emission and reflection on all bounding surfaces. Therefore, the zero radiative flux condition imposed on the symmetry planes would be equivalent to an erroneous assumption that the ray reaching these imaginary planes can be reflected therefrom in an arbitrary direction. In fact the plane of symmetry behaves as an ideal mirror. This can be explained by simple reasoning (see Fig 6.13). Consider a ray originating at point $\mathbf{r}$ and impinging the plane of symmetry at point $\mathbf{p}$. At this very point the ray leaves the subdomain of interest but simultaneously, another ray emitted at a point $\mathcal{M}\{\mathbf{r}\}$ being a symmetric image of point **r** enters the subdomain. Apart from its direction the ray entering the region has exactly the same features as that which leaves it. The direction of the incoming ray is symmetrical with respect to the normal to the plane of symmetry. Hence, this plane should be treated as if it were an ideal mirror. BEM offers an elegant method of treating symmetries of the sought for fields (see Section 5.6). An important feature of this approach, when applied to heat radiation, is that it does not introduce the assumption of diffuse reflection on planes of symmetry. Thus, this technique, originally developed to be employed in heat conduction problems, can be used to handle symmetry in BEM radiation codes [51] without any changes.

Presence of real mirrorlike reflecting surfaces.
As already shown, imaginary planes of symmetry behave like ideal mirrors. Obviously, the method used to handle the symmetry can be used to model specular reflections from real surfaces without any difficulty [51]. Hence, BEM is capable of handling problems in cavities consisting of diffuse and mirrorlike surfaces.

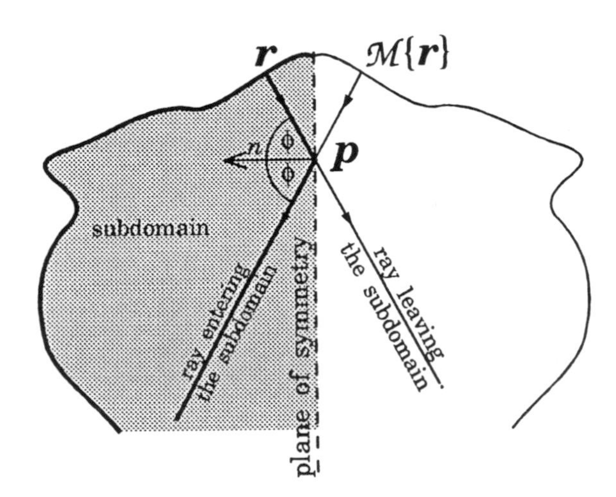

Figure 6.13: Rays emitted at point **r** and its image $\mathcal{M}\{\mathbf{r}\}$ crossing the plane of symmetry at point **p**.

Chapter 7

Other methods of solving heat radiation problems

7.1 Classification of solution techniques

The aim of this chapter is to overview the available methods of solving heat radiation problems, in the context of their links with the BEM technique. The stress will be laid on the mutual interrelations, strength and limitations of the reviewed methods. Their complete description is neither possible nor necessary here. Exhaustive discussions of the present methods of choice can be found in standard monographs on heat radiation [41, 64, 77, 82], reviews of heat radiation literature [43, 73, 93, 94, 95] and the references cited therein.

Due to the inherent complexity of all nontrivial heat radiation problems, they can be very seldom solved using analytical methods. The known analytical solutions are restricted mainly to 1D problems of minor interest in engineering. The solutions valid for black wall rectangular [74, 78] and cylindrical geometries [76] are of greater practical significance. Usage of analytical methods requires drastic simplifications of the physical model at hand. Therefore they can serve solely as a source of benchmark solutions to validate numerical techniques.

Known approximate methods of solving heat radiation problems fall into two classes:

i. methods based on introduction of simplifications into the physical formulation of the equation,

ii. methods using the complete transfer equations and solving these equations by means of approximate mathematical techniques.

In the first class of methods some terms of the transfer equation are neglected. Transparent gas (neglected attenuation and emission), cold medium (neglected emission), emission (neglected absorption), and diffusion (neglected far distance interactions) approximations are examples of frequently used techniques that belong to this group of methods. As in this book the attention is focused on mathematical techniques. This group of methods will not be addressed here.

The methods of the second group can belong to one of four groups:

- zoning methods based on the solution of the integral form of the equations of transfer in their standard (2.34); (2.41), or BEM (2.40), (2.42) formulations;
- methods based on the solution of the directional heat radiation equation (2.26);
- probabilistic methods;
- hybrid methods.

The BEM technique being the main topic of this book belongs to the first group of methods. Hence, in order to show the relationship of this approach to the classic zoning techniques, this group will be analyzed in more detail.

7.2 Zoning methods

Methods of this group have four common features:

- the domain of interest is subdivided into a finite number of subdomains (zones). Within zones the distributions of unknown fields are approximated by a set of simple functions;
- integral equations of radiative transfer are discretized;
- resulting equations are sets of algebraic equations;
- coefficients of these sets are computed by integration over zones.

The widely known *view factor method,* being the standard technique of solving radiative transport problems in cavities filled by transparent gas, is a typical example of the zoning approach. Another one is the well established *Hottel's zoning method,* a most frequently used engineering technique of handling 3D participating medium problems in enclosures. FEM [69, 70] and BEM techniques can be classified as zoning approaches. Also, frequently used engineering methods using course subdivision of the domain [49, 86] fall into this group of methods. Among many other variants of the zoning methods described in the literature the *product integration technique* [88] is the closest one to the BEM approach. Product integration is a general purpose method capable of solving multidimensional radiation problems in the presence of scattering. The technique utilizes, like BEM, piecewise polynomial approximation of the unknown functions and nodal collocation. Product integration relies on the discretization of the the classic integral equations and therefore it requires, as opposed to BEM, volume integration. Moreover, the concept of shape function approximation of the geometry is not employed in the product integration method. This means that this method is hard to implement in the case of 3D domains of complex shape (in the original paper [88] only 1D and 2D examples are solved).

Original derivation of both the view factor and Hottel's method [41] is based on physical reasoning. These techniques are described in textbooks on heat radiation and their standard derivation need not be reproduced here.

It has been recognized recently [10] that these methods can be interpreted as a specific version of the weighted residuals technique. This interpretation gives a deeper insight into the approximations behind these techniques. It also shows their common roots, and offers a simple means of improving their accuracy. Therefore this mathematical interpretation will be discussed briefly here.

7.2.1 View factor technique of solving transparent medium problems

Heat radiation in the presence of transparent medium is governed by Eq. (2.45). In order to discretize it the entire radiating surface S is subdivided into boundary elements ΔS_e; $1, 2, \ldots, E$.

The approximation of the unknown functions e_b and q^r is accomplished using constant shape functions defined as

$$\Phi^e(\mathbf{r}) = \begin{cases} 1 & \text{if } \mathbf{r} \in \Delta S_e \\ 0 & \text{if } \mathbf{r} \notin \Delta S_e \end{cases} \tag{7.1}$$

In terms of these shape functions the unknown functions can be approximated as

$$e_b[T(\mathbf{r})] \approx \sum_{e=1}^{E} e_b(T_e)\Phi^e(\mathbf{r}) \tag{7.2}$$

$$q^r(\mathbf{r}) \approx \sum_{e=1}^{E} q_e^r \Phi^e(\mathbf{r}) \tag{7.3}$$

where $e_b(T_e)$ and q_e^r stand for (constant within boundary element ΔS_e) values of blackbody emissive powers and radiative heat fluxes, respectively.

Equation (2.45) is discretized by the Galerkin technique using weighting and shape function being constant within elements (see Chapter 3). The result is a set of algebraic equations having the form

$$\mathbf{A}\mathbf{e_b}(T) = \mathbf{B}\mathbf{q}^r \tag{7.4}$$

with entries defined as

$$A_{ke} = -\int_{\Delta S_k}\int_{\Delta S_e} K(\mathbf{r},\mathbf{p})\, dS(\mathbf{r})\, dS(\mathbf{p}) + \delta_{ek} \tag{7.5}$$

$$B_{ke} = +\frac{1-\epsilon_e}{\epsilon_e}\int_{\Delta S_k}\int_{\Delta S_e} K(\mathbf{r},\mathbf{p})\, dS(\mathbf{r})\, dS(\mathbf{p}) - \frac{\delta_{ek}}{\epsilon_k} \tag{7.6}$$

where ϵ_e stands for the emissivity of the eth boundary element.

It should be noted that the general appearance of set (7.4) is the same as Eq. (6.7) obtained via BEM. However, to compute the entries of matrices **A** and **B** arising in Eq. (7.4) double integration should be carried out. This is in contrast with Eqs. (6.8) and (6.9) defining the entries of the BEM equation (6.7) where only single integration is required. On the other hand, the matrices obtained via the Galerkin technique are symmetric while their BEM analogs are not.

Double integrals arising in Eqs. (7.5) and (7.6) are called, in the literature, *configuration, view* or *angle factors.* Extensive tabulation of these factors for various geometry arrangements can be found in the literature [42, 77, 82]. The numerical values of these factors are typically obtained by analytical integration. The results are presented in the form of diagrams and formulae. The former are fairly cumbersome for computer implementation. Another disadvantage of this approach is that the number of arrangements that can be analyzed is limited to those that can be found in the literature.

The knowledge of the mathematical background of the view factor approach has two important consequences:

- the methodology of computing BEM matrix entries employing parametric representation of both geometry (see Section 5.2) and function can be readily adopted to compute view factors for arbitrary geometric arrangement. When adaptive integration (see Section 5.3.1) is used, the accuracy is limited only by the finite computer arithmetics;

- higher order approximation of unknown functions can be used. This leads to an increase in the accuracy of final results with only a slight increase of computer time.

Some works on applying FEM to solve heat radiation in transparent media has been reported in the literature [24, 45, 96]. The authors of the first reference used variational formulation of FEM. Remaining papers employ the weighted residuals technique of deriving FEM equations.

The first BEM formulations of heat radiation in transparent media were reported in Refs. [20, 46]. References [7, 46] seem to be the first where it was recognized that the view factor method can be interpreted in terms of the Galerkin technique.

7.2.2 Hottel's zoning method as a weighted residuals technique

Consider the set of integral equations (2.34) and (2.41), of heat radiation in participating media in their standard formulation. To discretize this set using weighting residuals, the surface of the enclosure is subdivided into a number of boundary elements ΔS_e exactly as described when discussing the view factor technique. Similarly, subdivision of the volume of the participating medium into a number of finite (volume) elements ΔV_j; $j = 1, 2, \ldots, L_V$, is introduced. Interpolating functions associated with this subdivision are step functions for both surface and volume elements. These sets of functions are defined as [see Eq. (5.3)]

$$\Phi^e(\mathbf{r}) = \begin{cases} 1 & \text{if } \mathbf{r} \in \Delta S_e \\ 0 & \text{if } \mathbf{r} \notin \Delta S_e \end{cases} \tag{7.7}$$

$$\Theta^j(\mathbf{r}') = \begin{cases} 1 & \text{if } \mathbf{r}' \in \Delta V_j \\ 0 & \text{if } \mathbf{r}' \notin \Delta V_j \end{cases} \tag{7.8}$$

where Φ and Θ stand for constant 2D and 3D shape functions, respectively.

There is a substantial difference between the volume (finite) elements introduced above and volume cells used in BEM. Volume elements are introduced to discretize volume integrals, while cells are used to discretize line integrals. Thus, 3D integration is performed over finite elements, and line integration is carried out along a ray going through a cell. From this viewpoint, Hottel's approach demands introduction of both volume cells and finite elements.

Following the standard weighting residuals procedure, the distributions of unknown functions (e_b, q^r, q^r_V) appearing in the integral equations to be discretized are expressed as linear combinations of the constant shape functions already defined, and nodal values of the sought for functions

$$e_b[T(\mathbf{r})] \approx \sum_{e=1}^{E} e_b(T_e)\Phi^e(\mathbf{r}) \tag{7.9}$$

$$q^r(\mathbf{r}) \approx \sum_{e=1}^{E} q^r_e \Phi^e(\mathbf{r}) \tag{7.10}$$

$$e_b[T^m(\mathbf{r}')] \approx \sum_{j=1}^{L_V} e_b[T^m_j]\Theta^j(\mathbf{r}') \tag{7.11}$$

$$q^r_V(\mathbf{r}') \approx \sum_{j=1}^{L_V} q^r_{Vj}\Theta^j(\mathbf{r}') \tag{7.12}$$

where T_e, T^m_j, q^r_e and q^r_{Vj} denote the nodal values of surface temperature, medium temperature, radiative heat flux, and radiative heat source, respectively. The nodes are placed in the centers of zones, i.e. boundary elements ΔS_e and volume elements ΔV_j.

Substitution of approximations (7.9)–(7.12) into integral equations (2.34) and (2.41) produces residuals. The final sets of algebraic equations are generated upon integrating a product of these residuals and elements of a sequence of weighting functions. These functions are chosen to be the same as the interpolating ones (7.7) and (7.8) which means that the integral equations are discretized using the Galerkin method.

Weighting Eq. (2.34) with constant surface element functions (7.7) and (2.41) with constant volume element functions (7.8) results in two sets of algebraic equations linking nodal emissive powers, radiative heat fluxes and radiative heat sources. The discretized versions of Eqs. (2.34) and (2.41) read

$$\left[\frac{q_k^r}{\epsilon_k} + e_b(T_k)\right] \Delta S_k = \sum_{e=1}^{E} \left[e_b(T_e) + \frac{1-\epsilon_e}{\epsilon_e} q_e^r\right] \overline{s_k s_e} + \sum_{j=1}^{L_V} e_b(T_j^m) \overline{s_k g_j} \tag{7.13}$$

$$[q_{Vi}^r + 4a_i e_b(T_i^m)] \Delta V_i = \sum_{e=1}^{E} \left[e_b(T_e) + \frac{1-\epsilon_e}{\epsilon_e} q_e^r\right] \overline{g_i s_e} + \sum_{j=1}^{L_V} e_b(T_j^m) \overline{g_i g_j} \tag{7.14}$$

In the above, for comparison purposes, the (somewhat strange) original Hottel's notation has been retained. The overbared double letter symbols can be interpreted as entries of matrices and are defined as

$$\overline{s_k s_e} = \int_{\Delta S_k} \int_{\Delta S_e} K(\mathbf{r}, \mathbf{p}) \tau(\mathbf{r}, \mathbf{p}) \, dS(\mathbf{r}) \, dS(\mathbf{p}) \tag{7.15}$$

$$\overline{s_k g_j} = a_j \int_{\Delta S_k} \int_{\Delta V_j} K_p(\mathbf{r}', \mathbf{p}) \tau(\mathbf{r}', \mathbf{p}) \, dV(\mathbf{r}') \, dS(\mathbf{p}) \tag{7.16}$$

$$\overline{g_i s_e} = a_i \int_{\Delta V_i} \int_{\Delta S_e} K_r(\mathbf{r}, \mathbf{p}') \tau(\mathbf{r}, \mathbf{p}') \, dS(\mathbf{r}) \, dV(\mathbf{p}') \tag{7.17}$$

$$\overline{g_i g_j} = a_i a_j \int_{\Delta V_i} \int_{\Delta V_j} K_0(\mathbf{r}', \mathbf{p}') \tau(\mathbf{r}', \mathbf{p}') \, dV(\mathbf{r}') \, dV(\mathbf{p}') \tag{7.18}$$

These entries are, in the literature, referred to as *surface-surface, gas-surface, surface-gas and gas-gas direct exchange areas*, respectively. Equations (7.13) through (7.18) are identical with those obtained by Hottel [41]. Thus, the classic zoning method can be interpreted as the Galerkin's weighted residuals solution of the standard integral equations of heat radiation (2.34) and (2.41) with weighting and approximating functions being constant in subregions. As the equations of the Finite Element Method are typically derived via the Galerkin technique, the zoning method has some common roots with FEM.

Knowing the mathematical interpretation of the zoning method, one can think of improving its accuracy by introducing higher order shape and weighting functions widely used in FEM. On the other hand, the weighted residuals technique is traditionally regarded as a purely numerical technique without any physical meaning. As previously mentioned the zoning equations were originally derived by physical reasoning [41]. Because the same result can be obtained using weighted residuals, some physical interpretation can be assigned to Galerkin's technique.

It can be also seen that the general form of the Galerkin sets of equations (7.13), (7.14) can be brought to the same general matrix form as those obtained by the BEM technique (6.42), (6.46). Similar to BEM, the entries of matrices are computed by integration over elements and when concave boundary elements are used some of the integrals are singular. In this case the no flux condition discussed in Section 6.5 can be employed to compute these integrals. Alternatively, when plane elements are used this condition can be employed to test the accuracy to which the entries of zonal matrices have been computed.

7.2.3 Comparison of Hottel's zoning method and BEM technique

Although the general form of the final algebraic equation sets of the zoning method and BEM are the same, the computer time required to form these sets differs significantly. Hottel's method leads to substantial problems in calculating the integrals, while these difficulties are much less when BEM formulation is used. To examine this a numerical example will be discussed.

Example 4
Accuracy of numerical integration in BEM and zoning methods

Heat radiation within a unit cube of gray walls dealt with in Example 2 (page 75) has been considered.

The problem has been solved using four approaches:

1. Hottel's method using quadratures of order 16 in one direction. The unknowns were the six radiative heat fluxes at walls 1 through 6;
2. Hottel's method using quadratures of order 16 in one direction. Additionally the surface integration was carried out upon subdividing the faces of the cube into four equal subsurfaces. The number of unknowns remains unchanged;
3. BEM using Gaussian quadratures of order four in one direction. The number of unknowns is equal to six;
4. BEM using the same Gaussian quadrature as in case 3. Four boundary elements were placed on each face of the cube. The number of unknowns was increased to 24.

It should be noted that to compute the entries of the zoning method by about $6 \cdot 16^6$ and $24 \cdot 16^6$, kernel function evaluations were needed for cases 1 and 2, respectively. BEM requires about $36 \cdot 16^1$ and $36 \cdot 16^2$ for cases 3 and 4, respectively. The relative error of satisfying the no flux condition (6.61) defined as

$$\delta = \frac{\sum_{j=1}^{N} A_{kj} - \sum_{j=1}^{L_V} C_{kj}}{\sum_{j=1}^{N} A_{kj}} \tag{7.19}$$

can be regarded as a measure of the accuracy of the numerical integration. The results are shown Table 7.1.

As can be seen from Table 7.1 the accuracy of calculating the entries of zoning matrices is, in case 1, very poor (error exceeds 300%). The results obtained by this method are therefore inaccurate. Other approaches produce comparable results. The reason for the unpleasant behaviour of the Hottel's method is the presence of

Table 7.1: Temperatures in K, radiative fluxes in Wm^{-2} and relative error of integration for various numerical techniques (Example 4).

Face	Temperature	Radiative fluxes			
		Numerical technique			
		1	2	3	4
1	300	29.91	138.9	156.7	144.8
2	1000	20.82	89.6	108.5	96.0[b]
3[a]	1000	17.43	91.7	108.8	97.5
4	1000	20.82	89.6	108.5	96.0[b]
5	1000	20.82	89.6	108.5	96.0[b]
6	1000	20.82	89.6	108.5	96.0[b]
Relative error		3.56	0.03	0.003	0.001
No of function evaluation		$6 \cdot 16^6$	$24 \cdot 16^6$	$36 \cdot 16^1$	$36 \cdot 16^2$

a wall lying opposite to face No 1
b mean value of four sub surfaces

a singularity of the integrand at the common edges of the faces. When integrating over one boundary element for a source point lying in the vicinity of that common edge, the integral becomes 'almost singular' and cannot be calculated accurately using classic quadratures.

The problem has been described in the literature and many efforts have been devoted to develop accurate algorithms of calculating the entries of the zoning matrices. To minimize the error of the numerically computed entries some smoothing and normalizing techniques have been suggested including regression [92] and Lagrangian multipliers [52]. Because of these difficulties the values of the entries have been computed using probabilistic quadratures [92] or measured [53]. The problem of accurate and efficient computation of the entries can be solved by the employing of adaptive integration discussed in Section 5.3.1.

Here are some factors influencing the efficiency of the compared BEM and Hottel's approaches:

- entries of zoning matrices are calculated by double integration, while BEM matrices are defined by single integrals;

- the kernel functions in both methods are strongly singular. As shown in Example 4 to reproduce the 'almost singular' behaviour of the integral, the zoning method requires considerably more function evaluations than BEM;

- no volume integration is required in BEM. Thus, the highest dimension of the integration domain that arises in BEM is 2D, while in the zoning method one needs to integrate over 6D domains (entry $\overline{g_i g_j}$);

- two types of integrals appears on the right hand side of BEM integral equations (2.40) and (2.42) and need to be discretized. The first integral is due to surface radiation and its attenuation in the medium. The second one is due to radiation emitted by the medium and attenuated within it. The discretization of the BEM integral equations is mainly connected with surface and line integral (transmissivity) discretization. This in turn is equivalent to calculation of the kernel functions K and K_r at quadrature nodes placed on boundary elements and the lengths d_l of rays intercepted by the boundaries of the volume cells. It will be shown that both quantities are determined when handling the first integral and can be used when discretizing the second one. To prove this it should be observed that the same subdivision into boundary elements is used when handling both integrals. From this observation it follows immediately that the nodes of quadratures employed to perform appropriate integration are, in the case of both integrals, identical. As the same kernel functions appear under the first and the second integral sign, the values of these functions at nodal points computed while discretizing the first integral can be, without any modification, used when discretizing the second one. Moreover, the line integrals (6.31) and (6.32) present in the definitions of the first and second integrals are computed by tracing rays connecting the same quadrature nodes. To compute both line integrals the lengths of rays d_l intersecting volume cells are to be determined. This is done when handling the first integral. The same values can be used to discretize the second one. Hence, in BEM the second integral can be discretized concurrently with the first one almost 'for free'. This is in contrast with Hottel's method where the second integral should be discretized independently from the first. Moreover, discretization of the second integral requires time consuming volume integration.

7.3 Methods employing the directional radiative transfer equation

7.3.1 General characteristics

This group of methods relies on solving the directional equation of transfer (2.26) along a given direction. The characteristic feature of this group of methods is that the working equations are differential ones, and because of this the methods are often referred to as *differential approximation methods.*

The main difficulty to be overcome when working with methods belonging to this group is the angular dependency of the radiation intensity. As pointed out in Chapter 2 intensity at a given point depends on the direction (angle) of ray propagation. Thus, an infinite number of intensity values can be assigned to a single point. To circumvent this problem the spatial and angular dependence of the intensity are separated. This is accomplished upon expressing the intensity as a sum of products of (unknown) spatially dependent coefficients A_j and (known) functions B_j depending solely on the solid angle Ω

$$I = \sum_{j=1}^{J} A_j(\mathbf{r})B_j(\Omega) \tag{7.20}$$

Functions $B_j(\Omega)$ approximating the angular dependence of the intensity can be chosen in an arbitrary manner. Types of these functions reported in the literature comprise constant, polynomials of direction cosines and spherical harmonics (transcendental functions relative to Legendre polynomials). Equation (7.20) is substituted into the directional equations of transfer (2.26) written for a chosen number of lines of sight. This results in a set of first order differential equations for the unknown coefficients A_j of the expansion (7.20). Integration with respect to the angular variable makes each equation of this set independent of direction. There are two methods of generating an appropriate number of equations, so that all coefficients A_j can be uniquely determined:

- subdivision of the entire solid angle into a number of intervals (some methods use overlapping intervals). To eliminate the angular variable approximation (7.20) is substituted into the directional equation of transfer and the result is integrated over the solid angle interval. In the next step the resulting set of differential equations is integrated along the center line of each interval. The number of intervals must be equal to the number of spatially dependent coefficients A_j of the approximation (7.20). Because integration of the intensity over the solid angle interval yields a one way radiative energy flux streaming along this direction this approach is usually termed the *flux method;*

- multiplying the original directional equation by subsequent powers of direction cosines and integrating the result over the entire solid angle 4π. This approach is known as the *moments method* .

It should be mentioned that the nomenclature of the directional radiation equation methods is not fixed. Also, some approaches combining features of more techniques can be found in the literature, e.g. approach using the moment equations and a subdivision of the entire solid angle into intervals [89]. In the remaining portion of the present section characteristic features of these methods in their standard formulations will be discussed briefly.

7.3.2 Flux method

The directional heat radiation equations integrated over the solid angle intervals are written for a chosen number of directions (usually along coordinate axes). This produces a set of ordinary differential equations of the first order for the one way radiative heat fluxes, whose directions coincide with the previously chosen directions. For each direction two one way energy fluxes are sought for thus, the number of equations is twice the number of the directions chosen. Each pair of the first order flux equations is combined to yield a second order partial differential equation. Hence the radiation equations can be solved using well established numerical techniques used to discretize energy and mass transfer equations which are ruled by partial differential equations. A generalization of all flux methods was reported in the literature [2].

In the classic formulation of the flux method the angular dependence of the intensity is assumed constant [35]. More advanced approaches use powers of the direction cosines [28, 55] to approximate the angular dependence of the intensity.

Another generalization of the flux method can be achieved upon subdividing the entire solid angle into a number of unequal intervals centered along directions coinciding with nodes of quadrature rules (usually of Gaussian or Lobato type). The intensity is assumed constant within each solid angle interval. The resulting set of first order differential equations is not, contrary to standard flux method, converted into corresponding sets of second order differential equations. Instead, integration of the first order equations is performed using conservative finite difference schemes. This approach is known as the *discrete ordinates method* [34, 44, 47, 91]. The sought for radiative fluxes and sources are computed upon integrating over the solid angle. Appropriate integrals are calculated by quadrature rules. As the directions are chosen in agreement with the nodes of such rules, numerical integration can be efficiently carried out.

7.3.3 Moments method

The functions arising in this method are termed *intensity moments* and are defined as integrals of the intensity and direction cosine products. Moments of intensity denoted as I_{ij}^{mn} are defined as

$$I_{ij}^{mn} = \int_{4\pi} I \left(l_{x_i}\right)^n \left(l_{x_j}\right)^m \, d\Omega \tag{7.21}$$

where l_{x_i} and l_{x_j} denote direction cosines with respect to coordinate axes x_i and x_j, respectively, and m, n stand for the powers to which the cosines are raised.

The first three moments have a certain physical interpretation. Zeroth moment $m = 0$, $n = 0$ is the entire radiative energy incoming at a given point. First moments $m = 1$, $n = 0$ are the x_i coordinates of the radiative flux vector, whereas the second moments $m = 1$, $n = 1$ are proportional to the radiative pressure tensor coordinates.

Some tedious algebra enables one to transform the set of first order differential equations of transfer into a set of second order partial differential equations that can be solved by well established methods. Polynomials [75] and spherical harmonics [26, 59, 68] are used to describe the angular dependence of the intensity. In the latter case the technique is termed *spherical harmonics method.*

7.3.4 Advantages and disadvantages

The methods developed in astrophysics, meteorology and nuclear science were usually for 1D problems. Generalization to 3D enclosure problems typically encountered in engineering shows many drawbacks of these techniques.

The first is that the initial conditions necessary to start the integration along the line of sight are not known explicitly. The angular variation of intensity, characteristic of multidimensional problems, causes that the initial condition (initial intensity) is dependent on solutions obtained when integrating along other directions. Practically, the initial intensity is obtained upon integrating over the hemisphere centered at the point of interest. Hence, to adjust the initial condition several iterations are required.

All methods belonging to the discussed group are based on a solution of differential equations. This is simultaneously the main drawback and the greatest advantage of these techniques. The advantage stems from the fact that the differential equations can be easily coupled with the convection, conduction and mass transfer equations within the participating medium. Hence, introducing a heat radiation option into an existing convective transfer code does not require much programming effort. It is also of importance that the same numerical grid can often be used to solve convection and radiation problems. On the other hand, differential equations cannot, by their nature, reproduce the far-distance interactions characteristic of radiative heat exchange. A proper treatment of the far-distance interactions can be achieved only by solving the integral equation formulation.

A measure of the radiation interaction between two points $\mathbf{r}$ and $\mathbf{p}$ is the *optical depth.* This quantity is defined as

$$\kappa = \int_0^{L_{rp}} a(\mathbf{r}')\, dL_{rp}(\mathbf{r}') \tag{7.22}$$

When the medium is *optically thick* the distance within which radiation emitted at one point influences another is small and the far-distance interactions can be neglected. In this situation the governing equation of radiative transfer can be approximated by a differential equation. This simplified treatment of radiative interchange is termed *diffusion approximation.* All methods based on the solution of the directional heat radiation equation utilize this approximation thus, they produce reasonable results for media of large optical depth (>0.5).

The coupling between solutions corresponding to different lines of sight is, as already mentioned, assured by the iterative procedure of adjusting the initial conditions. Additional coupling can be introduced by scattering (if present). In the situation when absorption and emission within a medium dominates over scattering, the differential approximation methods suffer from an additional error termed *ray effect* [34]. The source of the error is that the radiation is allowed to stream only along arbitrary chosen discrete directions. Radiation emitted at a point can remain 'unseen' by the observation point unless these two points lie along one of the chosen discrete directions. Scattering introduces additional coupling therefore it diminishes the ray effect. Increasing the number of directions along which the equation is integrated leads also to an improvement of the accuracy, but it mitigates the ray effect slowly.

7.4 Monte Carlo technique

7.4.1 Outline of the method

The technique is based on simulating a physical situation by sampling from distributions using (pseudo)random numbers. The distributions are chosen in such a way, that the result of the sampling reproduces the output of the physical model as close as possible, provided a sufficiently large number of trials have been carried out. French mathematician Buffon who, in 1768 determined the value of π number by casting a needle on a ruled grid, seems to be the first scientist who used this technique. The first modern applications to solve complex engineering problems are attributed to Stanisław Ulam and John von Neuman and their team involved in the research on nuclear weapons.

The first step of the Monte Carlo method, in its applications to radiative heat transfer, is the subdivision of the boundary and the domain in a finite number of isothermal elements (zones). Additionally, when nongray bodies are involved, the entire spectrum is subdivided into a number of intervals. Obviously, this step of the Monte Carlo technique is deterministic, and as a matter of fact it is the discretization of geometry and spectrum encountered in all zoning methods.

The second step of the Monte Carlo method, typically very time consuming, comprises of a loop running over all zones introduced in the first step.

For each surface and volume zone a bundle of energy is randomly generated. The wavelength, direction and coordinates of the point of the bundle origin are obtained by random sampling. The direction in which the bundle is emitted depends on the directional characteristic of the material where the energy was 'born'.

The emitted energy portion is traced on its way through the medium where it can be both scattered and attenuated, the probability of these events being a function of material properties within a given zone. Attenuation results in diminishing the energy of the bundle, while scattering causes a change of direction of the ray.

Bundles that reach solid walls can be either absorbed at the wall or reflected therefrom, the probability of these events depends on the reflexivity of the wall. In the case of diffusive reflectors, direction of the reflected ray is generated randomly with uniform probability, i.e. the probability that the ray is reflected in each direction is the same. Since the direction of a specular reflection is uniquely determined by the direction of the incoming ray, this model of reflection does not require random numbers generation. Arbitrary model of reflection can be readily accounted for.

A reflected ray can undergo attenuation and scattering before it reaches the solid walls. The procedure of tracing the bundle is continued until the bundle is absorbed or its remaining energy is negligible.

The fractions of bundles emitted from a given zone that have been absorbed by subsequent zones, are stored in a square matrix. The energy flux transmitted from the ith zone to the jth surface is calculated from a relationship (for simplicity gray body analysis is considered)

$$E_{ij} = \epsilon_i \, e_{bi} \, S_i \, p_{ij} \tag{7.23}$$

where E_{ij} denotes the portion of the entire energy emitted from the ith zone that was absorbed by the jth one. ϵ_i, e_{bi}, S_i are the emissivity, blackbody emissive power and surface of the ith zone respectively, while p_{ij} stands for the fraction of bundles emitted by the zone i that were absorbed by zone j.

The net energy gained by the jth element is computed as a sum of contributions E_{ij} due to emission of all elements i minus the energy emitted from the jth element.

For the sake of clarity the most straightforward version of the Monte Carlo technique has been discussed here. Many improvements of this method, leading to substantial computation time savings and increases in accuracy, have been developed. The *irradiation factor method* [71] and the *exodus method*, enabling one to avoid random number generation [31], are typical examples of such improvements. The detailed analysis of more advanced topics connected with Monte Carlo techniques is out of the scope of this book. A comprehensive description of these methods, along with an extensive review of literature dealing with the application of the Monte Carlo method to heat radiation, is available in Ref. [38]. A brief overview of the method can also be found in standard monographs [77]. Accuracy and convergence of probabilistic methods are addressed in Ref. [58]. The reader who wishes to extend his knowledge on Monte Carlo techniques is referred to these texts.

7.4.2 Zoning methods versus Monte Carlo

The main advantage of the Monte Carlo method is the relative ease with which various physical complexities can be accounted for. The mathematical background needed to write a Monte Carlo code is indeed very limited. The possibility of working with physical intuition rather than with sophisticated mathematical tools, is regarded by the majority of engineers as an advantage of this technique. The main difficulty when employing the Monte Carlo method in practice is the long computation time needed by codes based on this technique. When comparing such codes with computer implementations of the zoning method the following aspects can be taken into consideration:

- both methods use discretization of boundary, volume and, if appropriate, also the spectrum;

- the sampling process characteristic of Monte Carlo requires much additional computing time and introduces additional error;

- fractions p_{ij} of Eq. (7.23) are similar to entries of the matrices **A**, **B**, **C** of Eq. (6.42). One entry of these matrices, as well as the fractions p_{ij}, are computed upon tracing rays originating at one zone and impinging another. The rays traced in the zoning method are chosen to connect quadrature nodes placed on these two elements. These nodes, in turn, are chosen so as to minimize the integration error. In contrast with this, the rays traced in the Monte Carlo technique are selected by random. Moreover, in the zoning methods the rays are traced only until they impinge the solid wall for the first time. In the Monte

Carlo approach the process of ray tracing is much longer, as it is continued until the energy of the ray is small enough. This typically comprises several ray reflections from solid walls. All these features mean that to achieve an accuracy comparable with the zoning method, much more rays should be (longer) traced;

- when all temperatures in the system are known, the Monte Carlo technique does not require solving systems of equations. This is not a serious drawback of the zoning method because typically the time needed to solve the equations is much smaller than that of forming the matrices. Moreover, other boundary conditions (prescribed radiative heat fluxes or sources) cannot be handled easily using this technique;

- some problems involving very complex physical situations such as nondiffuse reflections, anisotropic scattering cannot be solved by zoning methods in a straightforward manner. Modelling such situations by Monte Carlo techniques presents no significant difficulties.

Generally, the use of the Monte Carlo approach is advisable in complex situations that cannot be handled using the zoning approach. Another important area of successful application of the Monte Carlo method is radiation in very complex (or random) geometry, e.g. bulk materials or fluidized beds. The popularity of the Monte Carlo technique is attributed to the ease of coding these algorithm. Monte Carlo can also be useful in producing benchmarks to validate other approaches. It should be kept in mind that to run a Monte Carlo code access to a powerful computer is essential.

7.5 Hybrid methods

The group of techniques comprises approaches possessing features characteristic of different groups of solution methods of the radiation interchange problem. One of such approaches, containing elements of the zoning and probabilistic method, has been already mentioned on page 100. In this approach [92] the entries of the zoning matrices were computed via probabilistic quadratures. This approach leads to excessive computing times and cannot be recommended.

Another hybrid technique is the *discrete transfer method* [56]. The approach uses zonal discretization of geometry with isothermal surface and volume elements. Rays impinging points placed at the center of each surface element are traced. The directions of these rays are chosen so as to coincide with nodes of simple numerical quadratures over the solid angle, similarly as in the discrete transfer method. While tracing the rays, integration along the line of sight is performed analytically, assuming constant temperature and absorption coefficient within a given cell, similarly as in the zoning method. However, the intensities at the starting points (initial conditions for ray tracing) of these rays are unknown, because they depend on the unknown radiation incoming at these points. Hence, to adjust the initial conditions, an iterative procedure is needed. The calculation of the radiative heat sources within volume zones is, in this method, accomplished by a control volume approach relying on accumulation of the raise (lost) of the intensity for each ray passing a given element. The method links features of the zoning and directional equation methods. The main disadvantage is that it demands iterations even if all temperatures are given. Some authors claim [95] that the discrete transfer method is also prone to the ray effect error.

Chapter 8

Coupling radiation with other modes of heat transfer

8.1 Possible interactions between radiation, conduction and convection

Radiative heat transfer and other modes of heat exchange are usually mutually interrelated. Two types of interactions arise when such a coupling is taking place:

- radiation contributes to boundary conditions of heat conduction and/or heat convection. This contribution is termed the radiative heat flux and denoted as q^r. The notion of radiative heat flux has been introduced in Section 2.1.2 and defined as a net heat flux gained by an infinitesimal surface as a result of radiation. This type of interaction is characteristic of radiation in cavities and enclosures formed by solid, heat conducting walls;

- radiation contributes to the volumetric heat source term in the governing equation of conduction or convection. The additional heat generation rate caused by radiation is referred to as the radiative heat source and denoted as q_V^r. It has been defined in Section 2.1.2 as the heat gained as a result of radiative transfer by an infinitesimal volume during an elementary time increment. This type of interaction arises when heat is transferred in semitransparent media, e.g. glass, silicon, triatomic gases.

Two governing equations of radiation both in their standard [Eqs. (2.34) and (2.41)] and in BEM [Eqs. (2.40) and (2.42)] formulations link four unknown functions:

i. temperature of the boundary; $T(\mathbf{r})$,

ii. temperature of the participating medium; $T^m(\mathbf{r}')$,

iii. radiative heat flux on the boundary; $q^r(\mathbf{r})$,

iv. radiative heat source within the participating medium; $q_V^r(\mathbf{r}')$.

The discretized versions of the governing equations [Eqs. (6.42) and (6.46)] link vectors of nodal values of these functions. Hereafter only the matrix versions of governing equations will be used in the analyses.

As radiation provides two equations linking four functions, unless two of these functions are known, radiation problems cannot be solved uniquely. The question of a proper choice of the prescribed functions has been addressed in Section 6.6. The values of the two functions closing the set of equations can be found by carrying out appropriate measurements on existing objects. Another method to obtain a uniquely solvable set of equations is to append, to the equations of radiation, two independent equations. The first of these equations is the heat conduction one written for solid walls enclosing the radiating domain. The second equation is the energy conservation equation (conduction or convection) for the participating medium filling the enclosure. In the remaining portion of this chapter some specific situations arising in practice will be examined.

8.2 Transparent medium of known temperature filling a radiating enclosure

A steady state temperature field in a doubly-connected solid forming an enclosure (cavity) containing a transparent medium of known temperature is considered. The solid is bounded by two surfaces: internal—forming the enclosure, and external—bounding the entire solid. The internal surface, referred to as the radiating one, exchanges heat both by radiation with other portions of itself and by convection with the medium filling the enclosure. The external surface bounding the entire solid is subjected to linear boundary conditions of Dirichlet, Neuman, and Robin type. This case has been discussed in Ref.[20].

Because the medium filling the enclosure does not participate in the radiative heat transfer, the radiative heat sources need not be computed. Moreover, the temperature of the medium is known so that the number of unknown fields reduces to two. These are: the temperature within the solid (including the temperature of the internal surface) and the radiative heat fluxes on the surface of the enclosure. The equations to be solved for the problem under consideration are: the equation of conduction within the solid and the equation of radiation.

The coupling of these equations is ensured by the boundary condition on the internal, radiating wall. Condition (5.40) describes such a nonlinear boundary condition. Its discretized version can be written as

$$q_j^R = h_j(T_j^R - T_j^f) + q_j^r \tag{8.1}$$

where indices j correspond to the jth nodal point placed on the radiating surface. Superscript R has been introduced to distinguish nodes placed on the radiating surface from those laying on other boundaries.

The relevant heat conduction equation (5.36) has been derived in Section 5.4.2 and will be curtailed to the specific situation at hand

$$\mathbf{Kt} = \mathbf{f} + \mathbf{Dq}^r \tag{8.2}$$

The superscript n occurring in Eq. (5.36) has been changed to r because the only nonlinear (temperature dependent) heat flux q^n is the radiative one.

The vector of unknowns $\mathbf{t}$ can be split into two subvectors:

$\mathbf{T}^R$– vector containing unknown nodal temperatures of the radiating surface,

$\mathbf{t}^L$– vector containing unknown nodal temperatures and conductive heat fluxes of the nonradiating surfaces, referred to hereafter as linear unknowns vector.

Superscript L designates hereafter quantities associated with the external, nonradiating surface with linear boundary conditions prescribed.

For simplicity the gray model of diffusively radiating walls will be considered. The heat radiation equation has, in this specific, transparent medium case, an appearance [see Eq. (6.7)]

$$\mathbf{A}\sigma(\mathbf{T}^R)^4 = \mathbf{Bq}^r \tag{8.3}$$

In the above the Stefan-Boltzmann law (2.14) has been used to replace the emissive powers by fourth powers of nodal temperatures. $(\mathbf{T}^R)^4$ denotes the vector containing fourth powers of the nodal temperatures at nodes placed on the radiating boundary.

Equations (8.2), (8.3) constitute a set of two algebraic equations for the unknown nodal temperatures $\mathbf{T}^R$ and radiative heat fluxes $\mathbf{q}^r$ at nodes placed on the radiating boundaries, and unknowns stored in vector $\mathbf{t}^L$. The latter are associated with nodes placed on the nonradiating portion of the boundary. The elements of vector $\mathbf{t}^L$ are temperatures or conductive heat fluxes, depending on the boundary condition imposed at a given node.

The presence of fourth powers of the temperatures makes the coupled sets of Eqs. (8.2) and (8.3) nonlinear. Solution of these sets can be obtained using a general purpose nonlinear equation solver described elsewhere (see, e.g. [27, 90]).

The relationships constituting the set are linear in radiative heat fluxes $\mathbf{q}^r$ and unknowns associated with the nonradiating walls $\mathbf{t}^L$. Therefore, both unknown vectors can be pre-eliminated prior to entering the iterative loop. Such an approach leads to substantial computer time savings, because the set solved iteratively contains a minimum number of degrees of freedom.

The pre-elimination is carried out in two steps, each being a Gaussian condensation. In the first step set (8.2) is solved for the unknowns on the nonradiating surface $\mathbf{t}^L$, using standard Gaussian elimination. Let the subscript $G1$ denote the result of the (first) Gaussian pre-elimination. Then, the resulting set can be split into two subsets having an appearance (see Fig. 8.1)

$$\mathbf{K}^R_{G1}\mathbf{T}^R = \mathbf{f}^R_{G1} + \mathbf{D}^R_{G1}\mathbf{q}^r \tag{8.4}$$

$$\mathbf{t}^L = -\mathbf{K}^L_{G1}\mathbf{T}^R + \mathbf{f}^L_{G1} + \mathbf{D}^L_{G1}\mathbf{q}^r \tag{8.5}$$

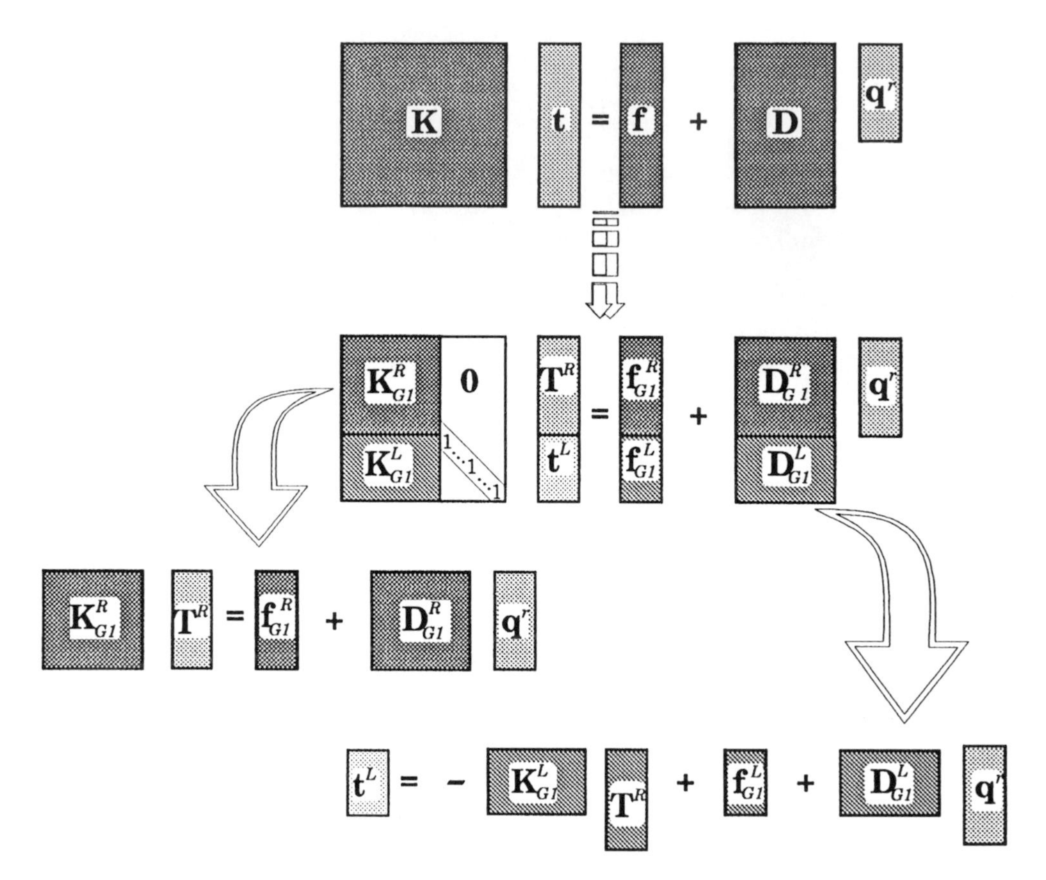

Figure 8.1: Pre-elimination of linear unknowns from equation of conduction (8.2).

Equations (8.3) and (8.4) are linear in radiative heat fluxes $\mathbf{q}^r$, hence they can be solved for this vector. Marking the result of the second Gaussian pre-elimination with the subscript $G2$ these equations can be rewritten as (see Fig. 8.2)

$$\mathbf{q}^r = \mathbf{A}_{G2}\sigma(\mathbf{T}^R)^4 \tag{8.6}$$

$$\mathbf{q}^r = \mathbf{K}^R_{G2}\mathbf{T}^R - \mathbf{f}^R_{G2} \tag{8.7}$$

Substitution of (8.6) into (8.7) yields a nonlinear set of equations for temperatures of the radiating surface

$$\mathbf{A}_{G2}\sigma(\mathbf{T}^R)^4 - \mathbf{K}^R_{G2}\mathbf{T}^R = -\mathbf{f}^R_{G2} \tag{8.8}$$

The solution of this set for the temperatures $\mathbf{T}^R$ of the radiating walls can be accomplished efficiently using the standard Newton-Raphson technique. Once this is completed, radiative heat fluxes can be computed from Eqs. (8.6) or (8.7). The remaining unknowns $\mathbf{t}^L$ corresponding to nodes placed on the nonradiating boundary can be computed from Eq. (8.5).

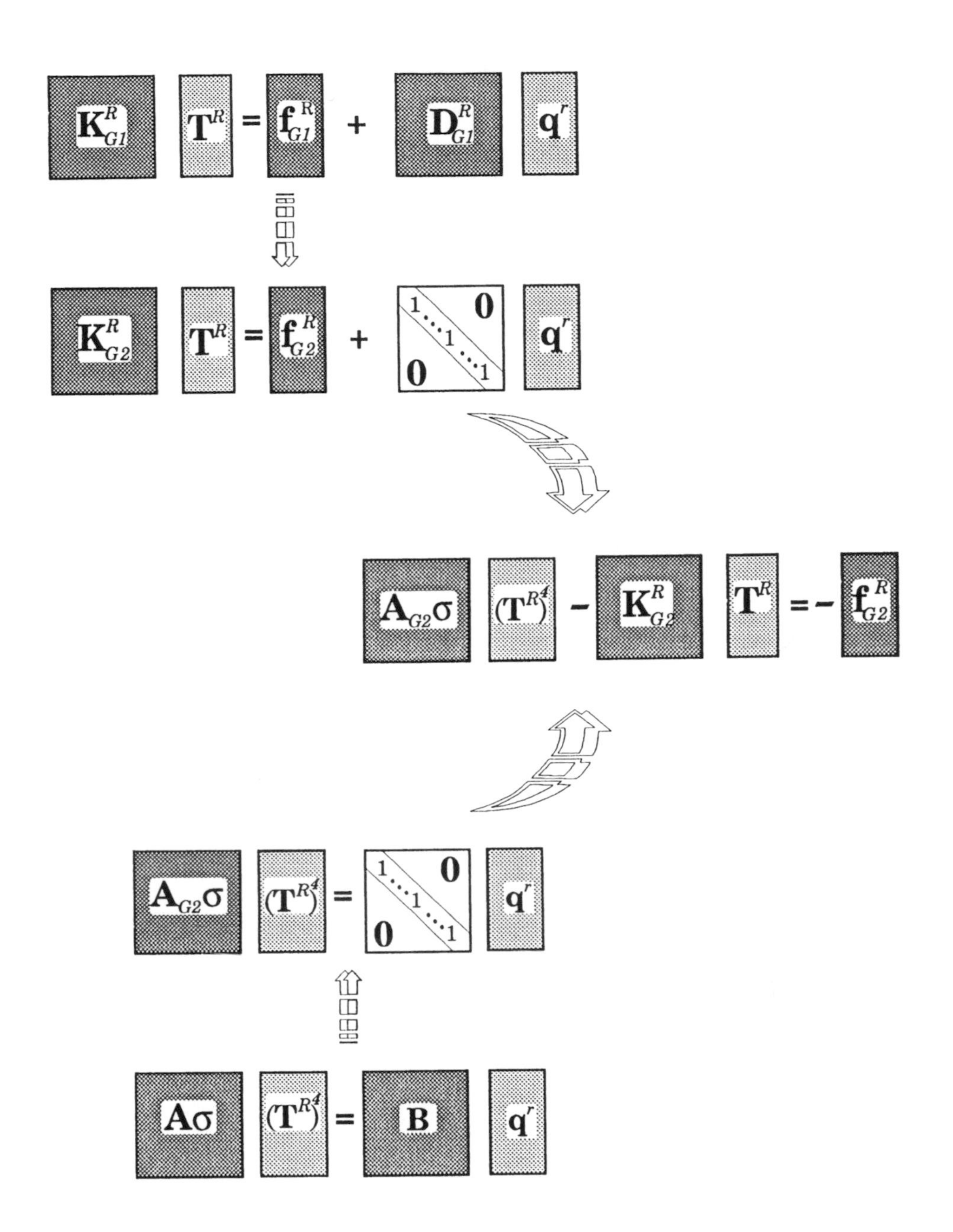

Figure 8.2: Pre-elimination of radiative heat fluxes from equation of radiation (8.3) and nonlinear equation of conduction (8.4).

8.3 Participating nonisothermal gray medium of known temperature filling a radiating enclosure

The problem under consideration is formulated similarly as that in the preceding section. The only difference is, that in place of a transparent medium filling the enclosure a gray participating one is considered. The problem discussed here has been addressed for a specific case of isothermal medium in Refs. [6, 11].

As the assumptions concerning conduction are the same as in Section 8.2 the equations of conduction (8.2) also apply in the problem at hand.

The allowance for participating medium requires a modification of the equation of radiation. Appropriate relationships (6.42) have been derived in Section 6.3 and will be recalled here with slight notation change

$$\mathbf{A}\sigma(\mathbf{T}^R)^4 = \mathbf{B}\mathbf{q}^r + \mathbf{C}\mathbf{e}_b(T^m) \tag{8.9}$$

Equations (8.2) and (8.9) form a set of coupled equations. They can be solved using general purpose nonlinear equation solvers. The Gaussian pre-elimination described in the preceding section can be adopted to deal with this slightly modified set of equations, without any difficulty. To employ this technique one should observe that the temperature distribution of the medium is known. Hence, the product of the matrix $\mathbf{C}$ and the vector of emissive powers $\mathbf{e}_b(T^m)$ can be reduced to a known vector. Denoting

$$\mathbf{c} = \sigma\,\mathbf{C}\,(\mathbf{T}^m)^4 \tag{8.10}$$

the equation of radiation can be rewritten in a form

$$\mathbf{A}\sigma(\mathbf{T}^R)^4 = \mathbf{B}\mathbf{q}^r + \mathbf{c} \tag{8.11}$$

Pre-elimination of the linear unknowns $\mathbf{t}^L$ from the equation of conduction (8.2) and elimination of the radiative heat fluxes from both the equation of radiation and conduction result in a set of nonlinear equations having a form

$$\mathbf{A}_{G2}\sigma(\mathbf{T}^R)^4 - \mathbf{K}^R_{G2}\mathbf{T}^R = -\mathbf{f}^R_{G2} + \mathbf{c}_{G2} \tag{8.12}$$

This set can be solved for unknown temperatures on the radiating boundary using the standard Newton-Raphson technique. Once this is completed the radiative heat fluxes and linear unknowns can be computed, similarly as in the preceding section, from relationships obtained as a result of pre-elimination.

To determine the radiative heat sources an additional effort is required. To calculate these sources, coefficients arising in Eq. (6.46) should be formed for each location of the point where heat sources are to be computed. Nodal values of the radiative heat sources can be then computed explicitly. The appropriate equation has, in matrix notation, an appearance

$$\mathbf{q}^r_V = \mathbf{L}\mathbf{q}^r - \mathbf{M}\sigma(\mathbf{T})^4 + \mathbf{N}\sigma(\mathbf{T}^m)^4 \tag{8.13}$$

where the matrices **L**, **M**, and **N** contain appropriate coefficients L_{ij}, M_{ij}, N_{il} defined in Section 6.3. The number of rows in Eq. (8.13) equals the number of points at which radiative heat sources are to be computed. One should notice that though Eq. (8.13) is given in a matrix form, it is explicit in radiative heat sources and hence, it need not be solved.

In Example 5 the influence of the participating medium onto the heat exchange in a thick walled radiating enclosure is examined.

Example 5

Temperature distribution in heat conducting walls forming an enclosure

A steady state 2D temperature field in a massive solid forming an enclosure is sought for. Within the enclosure heat convection and radiation is taking place. The enclosure is filled by an isothermal participating gas of known temperature $T^m = 1300K$. The heat transfer coefficient on the internal surface is $h^m = 20W/(m^2 \cdot K)$. A portion of the external surface is insulated whereas the remaining portion of this surface exchanges heat with a fluid of temperature $T^f = 290K$. The heat transfer coefficient is $h^f = 20W/(m^2 \cdot K)$. The emissivity of the internal surface is $\epsilon = 0.8$ and the heat conductivity of the solid is $k = 1.8W/(m \cdot K)$. The absorption coefficient of the medium is $a = 1m^{-1}$. The geometry and prescribed boundary conditions are depicted in Fig. 8.3. The distribution of temperature along the perimeter of the enclosure is shown in Fig. 8.4. For comparison, the temperature distributions corresponding to the case of negligible radiation ($\epsilon = 0$) and transparent medium ($a = 0$) are also shown in this figure. Figure 8.5 shows the calculated distribution of the radiative heat flux along the perimeter of the enclosure.

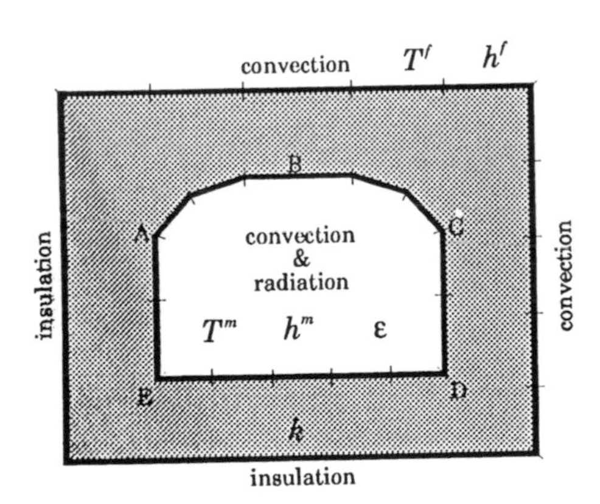

Figure 8.3: Geometry, numerical grid and boundary conditions prescribed. Example 5.

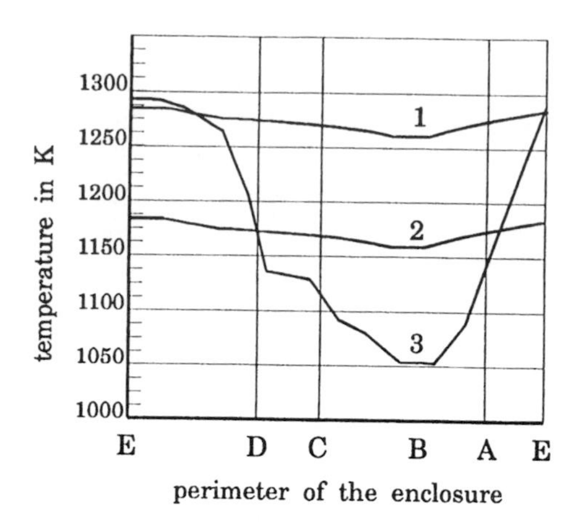

Figure 8.4: Temperature distribution along the perimeter of the enclosure. Example 5.
1) –participating medium $a = 1\,m^{-1}$,
2) –transparent medium $a = 0$,
3) –radiation neglected $\epsilon = 0$.

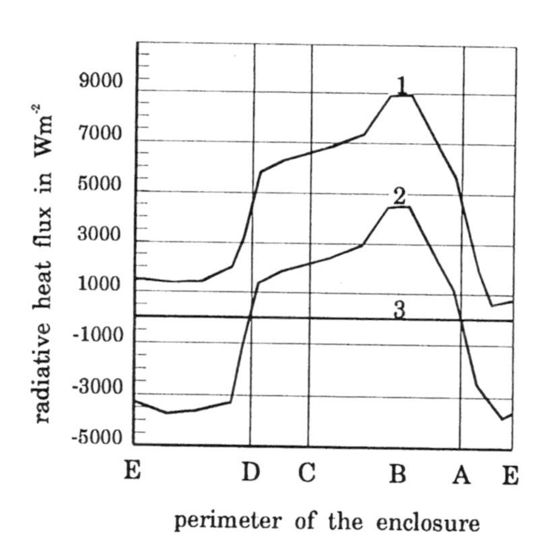

Figure 8.5: Radiative heat flux distribution along the perimeter of the enclosure. Example 5.
1) –participating medium $a = 1\,m^{-1}$,
2) –transparent medium $a = 0$,
3) –radiation neglected $\epsilon = 0$.

8.4 Participating nonisothermal gray medium of unknown temperature filling a radiating enclosure

This is the most general case. The unknown distributions to be determined in this case are: temperatures and radiative heat fluxes on the boundary of the solid forming the enclosure, radiative heat sources and temperature of the participating medium.

To solve the problem four equations are to be formulated. These are:

i. conduction within the solid walls forming the enclosure [Eq. (8.2)],

ii. radiative transfer within the participating medium [Eq.(8.9)],

iii. distribution of radiative heat sources in participating medium filling the enclosure [Eq.(8.13)],

iv. energy conservation in participating medium.

The first three equations have been used when solving problems with known temperature of the medium (see Section 8.3). The fourth equation should be introduced to close the system of equations. This additional equation can be formulated either as the heat conduction equation, when the medium filling the enclosure is a semi-transparent solid, (e.g. glass, silicon), or as the heat convection equation, when the participating medium is a fluid, (e.g. combustion gases, molten glass).

The second case results in very complex computations, as to solve the convection equations the velocity field should first be determined. This in turn can be accomplished upon solving momentum equations describing the fluid movement. The fluid flow of the participating medium can be imposed externally by a fan, a blower, or a pump. This mode of heat transfer is named *forced convection*. Another driving force of the fluid flow can be the buoyancy caused by density gradients of the fluid. This type of heat transfer is termed *free convection*. Solution of heat convection equations is, in both cases, a nontrivial task, especially when turbulent flows are involved. There is an immense amount of literature on numerical fluid mechanics and this question is out of the scope of the present book. The interested reader is referred to the handbook [60] and review paper [66] and the literature cited therein.

The general appearance of the equation of energy conservation to be included in the analysis can be written in a matrix form as

$$\mathbf{P}\mathbf{T}^R + \mathbf{R}\mathbf{q}^r + \mathbf{S}\mathbf{T}^m + \mathbf{U}(\mathbf{q}_V + \mathbf{q}_V^r) = 0 \tag{8.14}$$

where the entries of the matrices **P**, **R**, **S**, **U**, depend on the geometry, heat capacity, density and velocity fields. The specific form of these entries depends on the discretization method used when forming these matrices. Vectors $\mathbf{T}^R$, $\mathbf{q}^r$, $\mathbf{T}^m$, $\mathbf{q}_V^r$, $\mathbf{q}_V$ contain nodal values of the temperature of the enclosing surface, radiative heat fluxes on this surface, temperature of the medium, radiative heat sources and heat sources caused by phenomena other than radiation, respectively.

Simultaneous solution of the four coupled sets of equations mentioned above yields the sought for fields. The solution can be achieved using general purpose nonlinear equation solvers or schemes curtailed to the specific structure of the equations. Pre-elimination as discussed in previous sections is one of the possible choices.

One should keep in mind that in most practical situations these sets should be solved in conjunction with momentum equations. The solution is cumbersome and requires special numerical techniques and powerful computers. The main difficulties are due to the poor convergence of the iterative schemes employed to solve the momentum equations. The interested reader is referred to the aforementioned literature on heat convection to study this problem.

The described multi mode heat transfer models can be developed to deal with problems arising in many branches of science and technology. One of the most common applications is the mathematical modelling of the heat transfer within combustion chambers and industrial furnaces. Additional difficulties associated with such problems come from the presence of flames, with its complex chemical reactions and mass transfer phenomena. Recent literature on this difficult problem is briefly reviewed in Ref. [21]. The algorithm for solving heat transfer problems in a furnace chamber developed in this reference enables one to take into account the influence of flame and transient conduction in the heated solid.

Radiation in participating media normally induces convective movements of the fluid. It is interesting that the problem of stability of such movement has already been formulated in terms of integral equations and solved using fixed point iteration by Pogorzelski, a Polish mathematician in 1920 [67].

Chapter 9

Concluding remarks. Topics of future research

The Boundary Element Method is an efficient method for solving heat radiation problems. Its mathematical and numerical roots come from the standard BEM technique used in many branches of engineering computations. However, BEM techniques of solving heat radiation problems can also be interpreted as a further development and improvement of the classic Hottel's zoning method. It has also been recognized that both methods can be interpreted as a weighted residual solution of the same integral equations. Thus, both approaches stem from a common origin. Moreover, they have practically the same areas of applications. The superior numerical behaviour of BEM, when compared with Hottel's technique, can be attributed mainly to two features of the former method:

- conversion of the volume integrals into surface ones;
- using well established discretization techniques as used in other applications of BEM.

More specifically the computing time economy comes from following features of BEM:

i. multiple integrals need not be computed;

ii. no volume integrals are to be evaluated;

iii. the governing equation of heat radiation contains two integrals: the first due to surface and the second due to volume radiation. These two integrals are, in BEM, discretized concurrently using the same numerical mesh. Values needed to discretize the second integral are obtained when treating the first one. The cost of BEM discretization of the integral equation is equivalent to the integration of the first, surface integral. This is in contrast with Hottel's method where discretization of the mentioned two integrals must be performed independently. Thus, in Hottel's method the cost of discretization is the sum of costs associated with the first and second integral. Moreover, as the second integral is a volume one its discretization is by one order more expensive than the first one;

iv. the singular behaviour of the integral equation requires special numerical treatment. This problem is also encountered in standard BEM applications and appropriate BEM routines have some built in accuracy control and speed up mechanisms. The possibility of using the same routines in heat radiation frees the user from all difficulties associated with the treatment of the singularity. Long computing times and problems with achieving adequate accuracy when computing entries of Hottel's zoning matrices are the bottle necks of this method;

v. a large portion of BEM codes, e.g. those controlling the input of geometry, forming and assembling of matrices, and solvers can be used to handle radiation problems with only minor changes;

vi. the possibility of working with higher order approximations of both function and geometry.

Besides its computational advantages BEM also offers some conceptual ones. This especially concerns multimode heat transfer. Using BEM for the conductive and radiative portion of the analysis causes that all equations of the model are formulated as integral ones, and can be analyzed and solved using the same techniques. Other techniques work with integral equations (radiation) and partial differential ones (conduction and convection) each demanding different numerical treatment.

Applying BEM to solve heat radiation problems has a fairly short history. Thus, the experience in employing this technique to large scale problems fails. The main areas that demand further investigations are:

1. inclusion of various models of gas emission and absorption,

2. allowance for scattering,

3. allowance for nondiffuse emission and reflection of walls,

4. coupling with convection codes,

   4.1. laminar convection,

   4.2. convection with turbulence models,

   4.3. flame modelling,

   4.4. interaction of turbulence and radiation. It has been recognized that due to the nonlinear dependence of the emission on temperature its fluctuations caused by the turbulence strongly influence the radiation [80],

6. efficient shadow zone algorithms,

7. interaction with phase change (ablation, glass processing),

8. coupling with commercial heat conduction codes.

The numerical efficiency of BEM applied to the solution of heat radiation problems coupled with other modes of heat transfer make these areas of research promising.

Appendix A

Computing lengths of rays intersecting volume cells

The necessity of computing the lengths of rays d_l intersecting a volume cell arises in both BEM and classic zoning formulations, when the participating medium is nonisothermal and/or its absorption coefficient varies with position. The length of rays are required to compute approximate values of transmissivity [Eq. (6.33)] needed both in BEM and in the classic zone formulation. Additionally BEM requires these values to compute the value of line integral J_2 defined by Eq. (6.39). The time needed to determine the values of these lengths significantly influences the overall time of solving radiation problems in participating media. Optimization of the algorithm of finding the lengths is of great importance for practical computations.

Assume the entire volume of participating medium has been divided into a finite number of cells. The shape of cells can be arbitrary. However, in order to speed up the computations, it is convenient to generate the cells by intersecting the entire volume of the medium by planes parallel to the coordinate plane. Thus, the boundaries of each cell are either rectangles parallel to coordinate axes or, for cells adjacent to the bounding surface, appropriate boundary elements. It should be stressed that the surface bounding the enclosure can be of arbitrary shape, i.e. they can be curvilinear or planar and not necessarily parallel to coordinate axes (see Fig. A.1). Let the coordinates of the planes intersecting the volume be denoted as

$$\begin{aligned} \text{parallel to } x = 0\text{; planes } x &= x_{Vk} \quad k = 1, 2, \ldots, I_x \\ \text{parallel to } y = 0\text{; planes } y &= y_{Vk} \quad k = 1, 2, \ldots, I_y \\ \text{parallel to } z = 0\text{; planes } z &= z_{Vk} \quad k = 1, 2, \ldots, I_z \end{aligned} \tag{A.1}$$

(In Fig. A.1, $I_x = 6$, $I_y = 5$).

Consider a line of sight L_{rp} going from point $\mathbf{r}$ and reaching point $\mathbf{p}$. Let the Cartesian coordinates of these points be (r_x, r_y, r_z) and (p_x, p_y, p_z), respectively. The first step of the algorithm is to select from the set of plains defined by Eqs. (A.1) those which are intersected by the line L_{rp}. This can be accomplished by a simple search, as the planes of interest should satisfy the following inequalities:

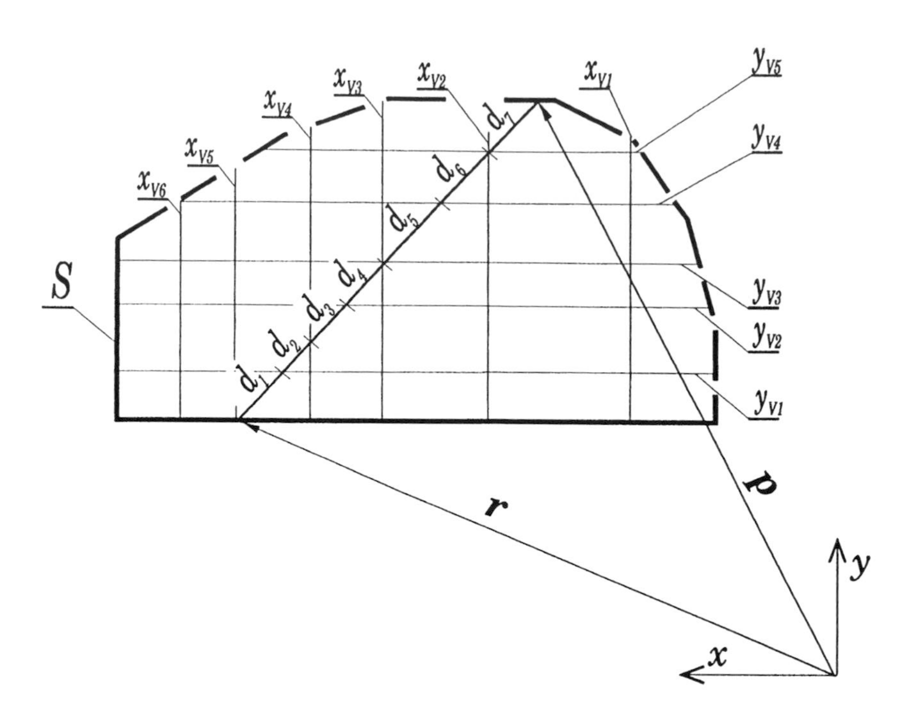

Figure A.1: Ray intersecting volume cells. Two-dimensional situation with the number of intersected cells $I_{rp} = 7$.

$$(x_{Vk} - r_x)(x_{Vk} - p_x) \; < \; 0 \tag{A.2}$$

$$(y_{Vk} - r_y)(y_{Vk} - p_y) \; < \; 0 \tag{A.3}$$

$$(z_{Vk} - r_z)(z_{Vk} - p_z) \; < \; 0 \tag{A.4}$$

Let $\widehat{x_{Vj}}; \quad j = 1, 2, \ldots, \widehat{I_x}$ denote the set of planes parallel to the $x = 0$ plane and satisfying inequality (A.2) (in Fig. A.1 these are planes x_{V4}, x_{V3}, and x_{V2}). The distance from point $\mathbf{r}$ to subsequent points of intersection of the line L_{rp} and the aforementioned set of planes can be calculated from the parametric equation of a straight line passing through points $\mathbf{r}$ and $\mathbf{p}$:

$$l_{xj} = |\mathbf{p} - \mathbf{r}| \frac{\widehat{x_{Vj}} - r_x}{p_x - r_x} \tag{A.5}$$

where $l_{xj}; \quad j = 1, 2, \ldots, \widehat{I_x}$ are the distances from point $\mathbf{r}$ to the subsequent intersection points of line L_{rp} and planes $\widehat{x_{Vj}}$.

The points of intersections of the line of sight L_{rp} and planes satisfying the relationships (A.3) and (A.4) can be found analogously. The corresponding sets of distances are denoted as $l_{yj}; \quad j = 1, 2, \ldots, \widehat{I_y}$ and $l_{zj}; \quad j = 1, 2, \ldots, \widehat{I_z}$ (in Fig. A.1 $\widehat{I_x} = 3$, $\widehat{I_y} = 5$.

The next step of the algorithm is to sort, in increasing order, the distances stored in l_{xj}, l_{yj}, and l_{zj}. Let I_{rp} stand for the number of of intercepted cells

$$I_{rp} = \widehat{I_x} + \widehat{I_y} + \widehat{I_x} \tag{A.6}$$

and L_l, $l = 1, 2, \ldots, I_{rp}$ be the set of distances resulting from such a sorting. The length of rays intercepted by the boundaries of the given cell are then calculated from Eq. (6.35) as

$$d_l = L_l - L_{l-1}; \quad l = 1, 2, \ldots, I_{rp} \tag{A.7}$$

with $L_0 = 0$.

To trace a ray on its way from point **r** to **p** the global numbers of subsequent volume cells passed by this ray should be known. To obtain this information some simple bookkeeping is required. Necessary data are stored in two additional arrays.

The first array contains the numbers of six (in 2D four) cells adjacent to a given cell (upper, lower, left, right, front, and behind neighbour). This array is somewhat similar to the connectivity matrix encountered in FEM. The difference is that the latter contains numbers of nodal points rather than the numbers of cells.

The second array needed to trace the ray is used to store flags indicating whether an intersection point comes from intercepting the ray by planes of constant x, y, or z coordinate (planes $\widehat{x_{Vk}}, \widehat{y_{Vk}}, \widehat{z_{Vk}}$).

Information contained in these two arrays enables one to sort out the global numbers of cells intersected by each ray. Knowing this, appropriate values of absorption coefficients and temperatures can be associated with the length of rays within cells. This in turn suffices to compute the values of line integrals in question.

Bibliography

[1] M. Abramowitz and I. A. Stegun. *Handbook of Mathematical Functions.* Dover Publications, New York, 1965.

[2] M. N. Abramzon and F. N. Lisin. Flow methods for the radiant transfer equations in cylindrical geometry. *High Temperature Physics,* translated from *Teplofizika Vysokikh Temperatur*, **22**, 95–100, 1984.

[3] J. P. S. Azevedo and L. C. Wrobel. Nonlinear heat conduction in composite bodies: A boundary element formulation. *International Journal for Numerical Methods in Engineering*, **26**, 19–38, 1988.

[4] P. K. Banerjee and R. Butterfield. *Boundary Element Methods in Engineering Science.* McGraw Hill, London, 1981.

[5] H. Bartelds. Development and verification of radiation of radiation models. Combustor Modelling AGARD Conference Preprint No 275, Advisory Group for Aerospace Research & Development, North Atlantic Treaty Organization, 7 Rue Ancelle 92200 Neuilly sur Seine, France, 1979.

[6] R. A. Białecki. Applying BEM to calculations of temperature field in bodies containing radiating enclosures. In C. A. Brebbia and G. Maier, editors, *Boundary Elements VII, Vol.2*, pages 2–35 through 2–50. Springer-Verlag, Berlin and New York, 1985.

[7] R. A. Białecki. Heat transfer in cavities: BEM solution. In C. A. Brebbia, editor, *Boundary Elements X, Vol.2,* pages 246–256. Springer-Verlag, Berlin and New York, 1988.

[8] R. A. Białecki. Modelling 3D band thermal radiation in cavities using BEM. In C. A. Brebbia and J. J. Connor, editors, *Advances in Boundary Elements, Vol. 2: Field and Flow Solutions,* pages 116–135. Springer-Verlag, Berlin and New York, 1989.

[9] R. A. Białecki. Solving 3D heat radiation problems in cavities filled by a participating non-gray medium using BEM. In L. C. Wrobel, C. A. Brebbia and A. J. Nowak, editors, *Computational Methods in Heat Transfer, Proceedings of the First International Conference on Innovative Numerical Techniques in Heat Transfer, Vol. 2,* pages 205–225. Springer-Verlag, Berlin and New York, 1990.

[10] R. A. Białecki. Applying the boundary element method to the solution of heat radiation problems in cavities filled by a nongray emitting-absorbing medium. *Numerical Heat Transfer, Part A*, **20**, 41–64, 1991.

[11] R. A. Białecki. Solving coupled heat radiation-conduction problems using the boundary element method. *Zeitschrift für Angewandte Mathematik und Mechanik*, **71**(6), T596–T599, 1991.

[12] R. A. Białecki. Boundary element calculations of the radiative heat sources. In L. C. Wrobel, C. A. Brebbia and A. J. Nowak, editors, *Advanced Computational Methods in Heat Transfer II, Vol. 1: Conduction, Radiation and Phase Change; Proceedings of the Second International Conference on Innovative Numerical Techniques in Heat Transfer,* pages 205–217. Elsevier Applied Science, London 1992.

[13] R. A. Białecki. Solving nonlinear heat transfer problems using the boundary element method. In L. C. Wrobel and C. A. Brebbia, editors, *Boundary Element Methods in Heat Transfer*, International Series in Computational Engineering, pages 87–122. Elsevier Applied Science, London, 1992.

[14] R. A. Białecki and G. Kuhn. Upgrading BETTI: minimum distance between a source point and a boundary element and contact resistance boundary condition. Part 1 of a report for Mercedes Benz Co, contract 019 0653 111 49, Chair of Technical Mechanics, University of Erlangen Nürnberg, Egerlandstraße 5, D-8520, Germany, February 1991, (unpublished).

[15] R. A. Białecki and G. Kuhn. Boundary element solution of heat conduction problems in multizone bodies of nonlinear material. *International Journal for Numerical Methods in Engineering*, **35**(5), 799–809, 1992.

[16] R. A. Białecki and G. Kuhn. Boundary element solution of nonlinear material heat conduction problems with contact resistance. *Zeitschrift für Angewandte Mathematik und Mechanik*, **72**(6), T486–T471, 1992.

[17] R. A. Białecki and R. Nahlik. Nonlinear equations solver for large equation sets arising when using BEM in inhomogeneous regions of nonlinear materials. In C. A. Brebbia, W. Wendland and G. Kuhn, editors, *Boundary Elements IX, Vol. 1,* pages 505–518. Springer-Verlag, Berlin and New York, 1987.

[18] R. A. Białecki and R. Nahlik. Solving nonlinear steady state potential problems in inhomogeneous bodies using the boundary element method. *Numerical Heat Transfer, Part B*, **16**, 79–96, 1989.

[19] R. A. Białecki and A. J. Nowak. Boundary value problems for nonlinear material and nonlinear boundary conditions. *Applied Mathematical Modelling*, **5**, 417–421, 1981.

[20] R. A. Białecki, A. J. Nowak and R. Nahlik. Temperature field in a solid forming an enclosure where heat transfer by convection and radiation is taking place. In *Proceedings of the 1st National UK Heat Transfer Conference*, Leeds, pages 989–1000. Pergamon Press, London, 1984.

[21] R. A. Białecki and R. Weber. Heat transfer in industrial furnaces. Desk study IFRF Doc No G 00/y/4, International Flame Research Foundation, 1970 CA IJmuiden P.O. Box 10.000, The Netherlands, March 1991.

[22] C. A. Brebbia and J. Dominguez. *Boundary Elements - An Introductory Course.* McGraw Hill, London, 1988.

[23] C. A. Brebbia, J. C. F. Telles, and L. C. Wrobel. *Boundary Element Techniques: Theory and Applications in Engineering.* Springer-Verlag, Berlin and New York, 1984.

[24] G. Breitbach, J. Altes and M. Sczimarowsky. Solution of radiative problems using variational based finite element method. *International Journal for Numerical Methods in Engineering*, **29**, 1701–1714, 1990.

[25] R. D. Cess and S. N. Tiwari. Infrared radiative energy transfer in gases. In J. P. Hartnett and Jr. Th. F. Irvine, editors, *Advances in Heat Transfer, Vol. 8*, pages 229–283. Academic Press, New York, 1972.

[26] P. Cheng. Two-dimensional radiating gas flow by a moment method. *AIAA Journal*, **2**(9), 1662–1664, 1964.

[27] G. Dahlquist and Å. Björck. *Numerical Methods.* Prentice Hall, Englewood Cliffs, 1974.

[28] A. G. DeMarco and F. C. Lockwood. A new flux model for the calculation of radiation in furnaces. *La Rivista dei Combustibili*, **29**(5-6), 184–196, 1975.

[29] D. E. Edwards. Molecular band gas radiation. In T. F. Irwine and J. P. Hartnett, editors, *Advances in Heat Transfer, Vol. 16*, pages 115–193. Academic Press, New York, 1976.

[30] D. K. Edwards and A. Balakrishnan. Thermal radiation by combustion gases. *International Journal of Heat and Mass Transfer*, **16**(1), 181–188, 1973.

[31] A. F. Emery and W. W. Carson. A modification of the Monte Carlo method: the Exodus method. *Journal of Heat Transfer, Transactions of ASME*, **90**(3), 328–332, 1968.

[32] A. F. Emery, O. Johansson, M. Lobo and A. Abrous. A comparative study of methods for computing the diffuse radiation viewfactors for complex structures. *Journal of Heat Transfer, Transactions of ASME*, **113**(2), 413–422, 1991.

[33] B. A. Finlayson. *The Method of Weighted Residuals and Variational Principles.* Academic Press, New York, 1972.

[34] W. A. Fiveland. Discrete ordinates solution of the radiative transport equation for rectangular enclosures. *Journal of Heat Transfer, Transactions of ASME,* **106**(4), 699–705, 1984.

[35] A. D. Gosman and F. C. Lockwood. Incorporation of a flux model for radiation into a finite difference procedure for furnace calculations. In *Fourteenth Symposium (International) on Combustion*, pages 661–671. The Combustion Institute, Pittsburg, 1973.

[36] L. Grela. Using higher order shape functions in BEM applied to heat radiation (in Polish). MSc. thesis, Institute of Thermal Technology, Silesian Technical University, Konarskiego 22, PL 44-101 Gliwice, Poland, 1993.

[37] M. Guiggiani and A. Gigante. A general algorithm for multidimensional Cauchy principal value itegrals in the boundary element method. *Journal of Applied Mechanics, Transactions of ASME,* **112**, 906–915, 1990.

[38] A. Haji-Sheikh. Monte Carlo methods. In W. J. Minkowycz, E. M. Sparrow, R. H. Pletcher and G. E. Schneider, editors, *Handbook of Numerical Methods in Heat Transfer*, pages 672–722. Wiley Interscience, New York, 1988.

[39] D. Hearn and M. P. Baker. *Computer Graphics.* Prentice Hall, Englewood Cliffs, 1986.

[40] H. C. Hottel and E. S. Cohen. Radiant heat exchange in gas-filled enclosures: allowance for nonuniformity of gas temperature. *Transactions of the American Institution of Chemical Engineers,* **4**(1), 3–14, 1958.

[41] H. C. Hottel and A. F. Sarofim. *Radiative Transfer.* McGraw Hill, New York, second edition, 1973.

[42] J. R. Howell. *A Catalog of Configuration Factors.* McGraw Hill, New York, 1982.

[43] J. R. Howell. Thermal radiation in participating media: the past, the present and some futures. *Journal of Heat Transfer, Transactions of ASME,* **110**(4), 1220–1229, 1988.

[44] A. S. Jamaluddin and P. J. Smith. Predicting radiative transfer in axisymmetric cylindrical enclosures using the discrete ordinates method. *Combustion Science and Technology,* **62**, 173–186, 1988.

[45] W. E. Mason Jr. Finite element analysis of coupled heat conduction and enclosure radiation. In *Proceedings of the International Conference on Numerical Methods in Thermal Problems*, Swansea, Pineridge Press, 1979, Lawrence Livermore Laboratory preprint No UCRL-82857 attached to the proceedings.

[46] M. A. Keavey. A boundary integral solution for radiating enclosures. In *Boundary Element Methods, Theory and Applications*, pages 111–120. Meeting of the Stress Analysis Group of the Institute of Physics. London 1986.

[47] T. K. Kin, J. A. Menart and H.S. Lee. Nongray radiative gas analysis using the S-N discrete ordinates method. *Journal of Heat Transfer, Transactions of ASME*, **113**(4), 946–952, 1991.

[48] G. A. Korn and Th. A. Korn. *Mathematical Handbook for Scientists and Engineers.* McGraw Hill, New York, 1968.

[49] E. Kostowski. A mathematical model of heat transfer in the reverbatory furnace chambers of industrial furnaces, (in German). *Arch. Eisenhüttenwes.*, **45**, 381–387, 1974.

[50] G. Kuhn. Boundary element technique in elastostatics and linear fracture mechanics. In W. L. Wendland and E. Stein, editors, *Finite Element Method and Boundary Element Methods from the Mathematical and Engineering Point of View*, International Centre for Mechanical Sciences Courses and Lectures No. 301. Springer-Verlag, Wien and New York, 1988.

[51] G. Kuhn. Private communication, April 1990.

[52] M. E. Larsen and J. R. Howell. Least-squares smoothing of direct-exchange areas in zonal analysis. *Journal of Heat Transfer, Transactions of ASME*, **108**(1), 239–242, 1986.

[53] H. P. Liu and J. R. Howell. Measurements of radiation exchange factors. *Journal of Heat Transfer, Transactions of ASME*, **109**(2), 470–477, 1987.

[54] G. Löbel, W. Sichert and G. Kuhn. *BETTI Boundary Element Code for Heat Conduction*, (in German). Chair of Technical Mechanics, University Erlangen-Nürnberg, Germany, Lyoner Str. 18, 6000 Frankfurt/M 71, Forschungskuratorium Maschinenbau e.V. edition, 1987.

[55] F. C. Lockwood and N. G. Shah. Evaluation of an efficient radiation flux model for furnace prediction procedures. In R. W. Lewis, K. Morgan and W. G. Habashi, editors, *Proceedings of the VIth International Heat Transfer Conference,* Toronto 78, pages 33–38. National Research Council of Canada, Hemisphere, Washington D. C., 1978.

[56] F. C. Lockwood and N. G. Shah. A new radiation solution method for incorporation in general combustion prediction procedures. In *Eighteenth Symposium (International) on Combustion*, pages 1405–1414. The Combustion Institute, Pittsburg, 1981.

[57] G. Maier, G. Novati and P. Parreira. On boundary element elastic analysis in the presence of cyclic symmetry. In C. A. Brebbia, editor, *Boundary Elements VI*, pages 3–19 through 3–41. Springer-Verlag, Berlin and New York, 1984.

[58] J. D. Maltby and P. J. Burnd. Performance, accuracy, and convergence in a three-dimensional Monte Carlo radiative heat transfer simulation. *Numerical Heat Transfer, Part B*, **19**, 191–201, 1991.

[59] M. P. Mengüc and R. Viskanta. Radiative heat transfer in tree dimensional enclosures containing inhomogeneous, anisotropically scattering media. *Journal of Quantitative Spectroscopy and Radiative Transfer*, **33**(6), 533–549, 1985.

[60] W. J. Minkowycz, E. M. Sparrow, R. H. Pletcher and G. E. Schneider, Editors. *Handbook of Numerical Methods in Heat Transfer*. Wiley Interscience, New York, 1988.

[61] G. G. W. Mustoe. Advanced integration schemes over boundary elements and volume cells for two and three dimensional nonlinear analysis. In P. K. Banerjee and S. Mukherjee, editors, *Developments in Boundary Element Methods - 3*, pages 213–270. Elsevier Applied Science, London, 1984.

[62] A. J. Nowak. Solving coupled problems involving conduction, convection and thermal radiation. In L. C. Wrobel and C. A. Brebbia, editors, *Boundary Element Methods in Heat Transfer*, International Series in Computational Engineering, pages 145–173. Elsevier Applied Science, London, 1992.

[63] A. J. Nowak. Solving linear heat conduction problems by the multiple reciprocity method. In L. C. Wrobel and C. A. Brebbia, editors, *Boundary Element Methods in Heat Transfer*, International Series in Computational Engineering, pages 63–86. Elsevier Applied Science, London, 1992.

[64] M. N. Özişik. *Radiative Heat Transfer and Interactions with Conduction and Radiation*. John Wiley and Sons, New York, 1973.

[65] P. W. Partridge, C. A. Brebbia and L. C. Wrobel. *The Dual Reciprocity Boundary Element Method*. Elsevier Applied Science, London, 1991.

[66] S. V. Patankar. Recent developments in computational heat transfer. *Journal of Heat Transfer, Transactions of ASME*, **110**(4), 1037–1045, 1988.

[67] W. Pogorzelski. Equilibre d'une masse gazeusse rayonnante. *Sprawozdania i Prace Polskiego Towarzystwa Fizycznego*, **1**, 78–85, 1920–21.

[68] A. C. Ratzell III and J. R. Howell. Two-dimensional radiation in absorbing-emitting media using the P-N approximation. *Journal of Heat Transfer, Transactions of ASME*, **105**(2), 333–340, 1983.

[69] M. M. Razzaque, J. R. Howell and D. E. Klein. Coupled radiative and conductive heat transfer in a two-dimensional rectangular enclosure with gray participating media using finite elements. *Journal of Heat Transfer, Transactions of ASME*, **106**(3), 613–619, 1984.

[70] M. M. Razzaque, D. E. Klein and J. R. Howell. Finite element solution of radiative heat transfer in a two dimensional rectangular enclosure with gray participating media. *Journal of Heat Transfer, Transactions of ASME*, **105**(4), 993–996, 1983.

[71] Z. Rudnicki. Analysis of the radiative energy transfer in a nonisothermal gas filled enclosures using irradiation factors computed by the Monte Carlo method (in Polish). *Zeszyty Naukowe Politechniki Śląskiej*, **95**, 1986.

[72] A. Sala. *Radiant Properties of Materials.* Elsevier, Amsterdam, 1986.

[73] A. F. Sarofim. Radiative heat transfer in combustion: friend or foe? In *Twenty-First Symposium (International) on Combustion*, pages 1–23. The Combustion Institute, Philadelphia 1986.

[74] N. Selçuk. Exact solutions for radiative heat transfer in box shaped furnaces. *Journal of Heat Transfer; Transactions of ASME*, **107**(3), 648–655, 1985.

[75] N. Selçuk. Finite difference solution of three dimensional flux equations for radiative transfer in furnaces. In R. W. Lewis, K. Morgan and W. G. Habashi, editors, *Proceedings of the Vth International Conference on Numerical Methods in Heat Transfer; Vol. 5, Part 2;* Montreal, Canada, pages 907–917. Pineridge Press, Swansea, 1986.

[76] N. Selçuk and Z. Tahiroğlu. Exact numerical solutions for radiative heat transfer in cylindrical furnaces. *International Journal for Numerical Methods in Engineering*, **26**, 1201–1212, 1988.

[77] R. Siegel and J. R. Howell. *Thermal Radiation Heat Transfer.* Hemisphere, Washington D.C, second edition, 1981.

[78] R. Sigel. Analytical solution for boundary heat fluxes from a radiating rectangular medium. *Journal of Heat Transfer, Transactions of ASME*, **113**(1), 113–261, 1991.

[79] T. F. Smith, Z. F. Shen and J. N. Friedman. Evaluation of coefficients for the weighted sum of gray gases model. *Journal of Heat Transfer, Transactions of ASME*, **104**(4), 602–608, 1982.

[80] L. H. Song and R. Viskanta. Interaction of radiation with turbulence: application to combustion system. *Journal of Thermophysics*, **1**(1), 56–62, 1987.

[81] E. M. Sparrow, L. U. Alberts and E. R. G. Eckert. Thermal radiation characteristics of cylindrical enclosures. *Journal of Heat Transfer, Transactions of ASME*, **C84**(2), 73–81, 1962.

[82] E. M. Sparrow and R. D. Cess. *Radiation Heat Transfer.* Hemisphere, Washington D.C., second edition, 1978.

[83] F. R. Steward and Y. S. Kocaefe. Methods for determining the radiative transfer in a furnace chamber using fundamental equations of motion and transport of heat and mass. In J. Grigull, editor, *Proceedings of the 7th International Heat Transfer Conference*, pages 553–558. Hemisphere, Washington D.C., 1982.

[84] F. R. Steward and Y. S. Kocaefe. Total emissivity and absorptivity models for carbon dioxide and water vapor and their mixtures. In C. L. Tien, C. V. Carey and S. Ferrel, editors, *Proceedings of the 8th International Heat Transfer Conference*, pages 735–740. Hemisphere, Washington D.C., 1986.

[85] A. H. Stroud and P Secrest. *Gaussian Quadrature Formulas.* Prentice Hall, Englewood Cliffs, 1966.

[86] J. Szargut. Heat transfer by radiation in the chamber of the batch type furnace (in Polish). *Archiwum Hutnictwa*, **16**(2), 143–153, 1971.

[87] J. Szargut (Ed.), R. A. Białecki, K. Kurpisz, J. Skorek and Z. Rudnicki. *Numerical Modelling of Temperature Distribution (in Polish).* WNT, Warsaw, 1992.

[88] Z. Tan. Radiative heat transfer in multidimensional emitting absorbing and anisotropic scattering media-mathematical formulation and numerical method. *Journal of Heat Transfer, Transactions of ASME*, **111**(1), 141–147, 1989.

[89] Y. S. Chou and C. L. Tien. A modified moment method for radiative transfer in non-planar systems. *Journal of Quantitative Spectroscopy and Radiative Transfer*, **8**, 919–933, 1968.

[90] J. F. Traub. *Iterative Methods for the Solution of Equations.* Chelsea Publishing Company, New York, 1982.

[91] J. S. Truelove. Three dimensional radiation in absorbing-emitting-scattering media using the discrete ordinates approximation. *Journal of Quantitative Spectroscopy and Radiative Transfer*, **39**(1), 27–31, 1988.

[92] H. A. J. Vercammen and G. F. Froment. An improved zone method using Monte Carlo techniques for the simulation of radiation in industrial furnaces. *International Journal of Heat and Mass Transfer*, **23**(3), 329–337, 1980.

[93] R. Viskanta. Radiation heat transfer: interaction with conduction and convection and approximate methods in radiation. In *Proceedings of the 7th International Heat Transfer Conference*, Munich, pages 103–121. Hemisphere, Washington D.C., 1982.

[94] R. Viskanta. Radiative heat transfer. *Fortschritte der Vehrfahrenstechnik*, **22A**, 51–81, 1984.

[95] R. Viskanta and M. P. Mengüc. Radiation heat transfer in combustion systems. *Progress in Energy Combustion Science*, **13**, 97–160, 1987.

[96] L. D. Wills and H. Wolf. Evaluation of thermal and stress fields in a composite axisymmetric body with a nonlinear surface radiation boundary condition. In R. W. Lewis, K. Morgan, J.A. Johnson and R. Smith, editors, *Computational Techniques in Heat Transfer*, pages 71–94. Pineridge Press, Swansea, 1985.

[97] O. C. Zienkiewicz. *The Finite Element Method.* McGraw Hill, London, third edition, 1977.

Index

[1]boldface set pages refer to definitions of appropriate entries